W9-CKQ-584
Euclid Public Library
631 E. 222nd Street
Euclid, Ohio 44123
216-261-5300

IN THE
WOMB
Animals

IN THE WOMB

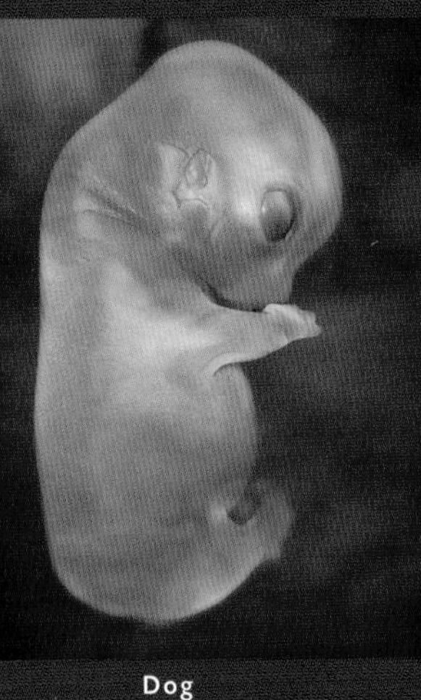
Dog

Kangaroo

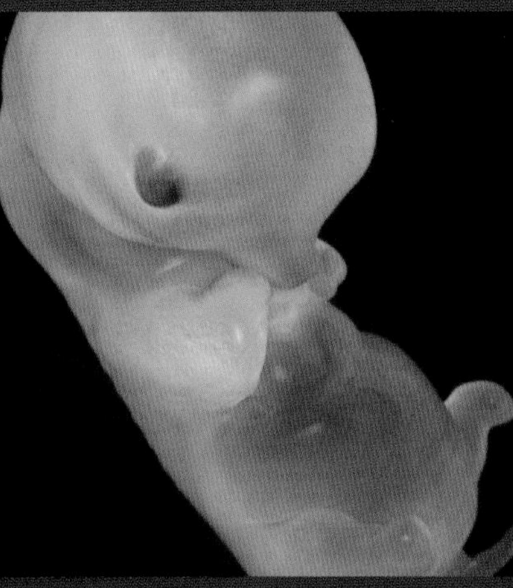
Elephant

Animals

Michael Sims

Dolphin

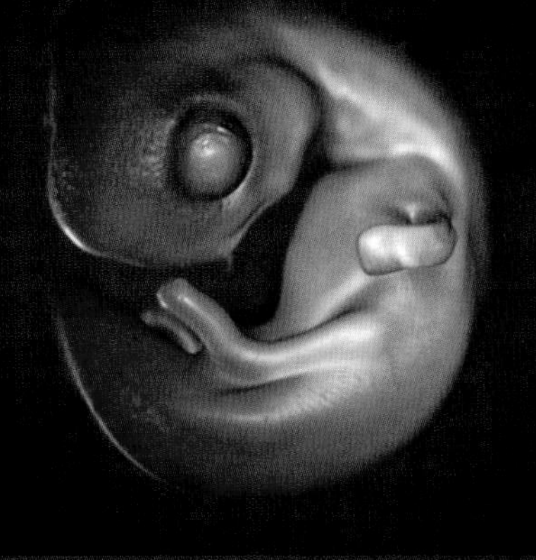

Penguin

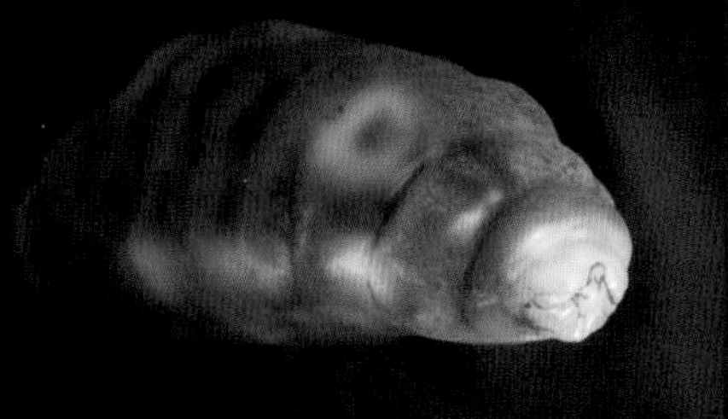

Wasp

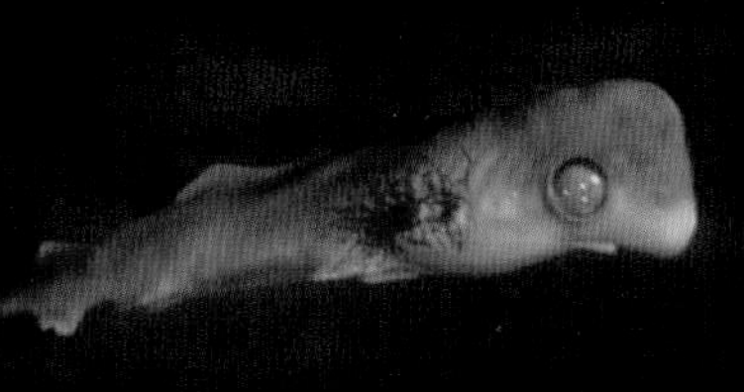

Shark

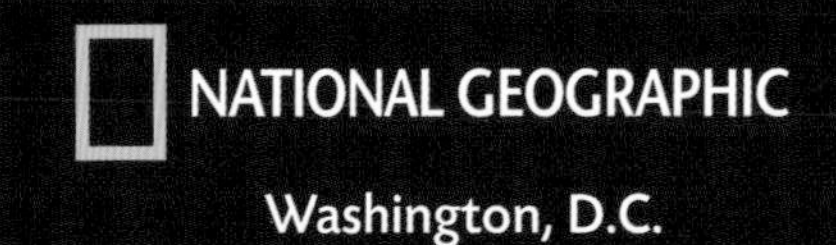

Contents

Introduction 6

Time Line 16

Part 1 Familiar 18

Part 2 Extraordinary 60

Part 3 Extreme 110

Afterword 154

Further Reading 156

About the Author 158

Credits 159

An Asian elephant fetus after 12 months in the womb

One Plus One Introduction

I grew up in the country in eastern Tennessee, raising dogs and rabbits and ducks, guinea pigs and hamsters, a turtle and a horse. My childhood was populated with a wonderful circus of animals, including whole galumphing litters of pups. I've kept a nature journal since I was a teenager, and in one of its initial entries I wrote about the birth of our dog's puppies. I already knew that most animals live danger-filled lives. On that day, however, because one pup was stillborn and another died soon afterward, I realized that the journey to birth was fraught with its own risks. Yet most of the pups not only survived but flourished—instinctively nursed, gradually opened their eyes, and soon gamboled across the lawn, exploring their new world.

So when an editor at National Geographic Books called and invited me to write a book about animals in the womb, how could I resist? I've watched the beautiful and mysterious process of birth from the outside so many times that I jumped at the chance to follow it from the inside. Learning how the world works, which is the beginning of caring about it, now and then requires that we gently violate the privacy of our nonhuman neighbors. When we do, we are reminded all over again how similar their lives are to our own.

This book is the second of the *In the Womb* series from National Geographic Books. The first, naturally, dealt with our favorite subject—ourselves—following the growth of a human being from conception to birth. Its inspiration, a documentary on the National Geographic Channel, begat a series of documentaries and now a series of books. Each explores various aspects of prenatal development in humans and other animals. The book you hold in your hand contains much original material, but I began it by weaving together strands of the research behind the documentaries *In the Womb: Animals* and *In the Womb: Extreme Animals*. Most of the remarkable illustrations from inside the womb came from the documentaries themselves, from their brilliantly realistic models and computer-generated imaging. It's exciting to watch scientific research, advanced technology, and the creative imagination unite in the goal of seeing the world more clearly.

Ladybug beetles mating

Felines, wild or tame, are both affectionate and fierce.

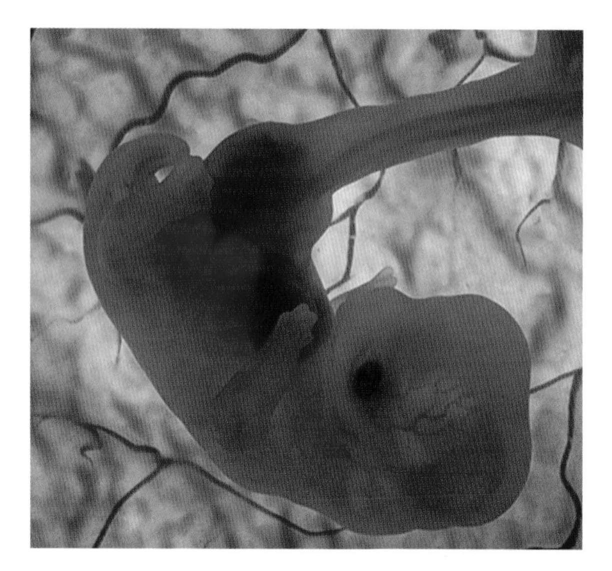

A dog embryo about 30 days after conception

In these pages, we follow the fascinating process of prenatal development through a trio of primary characters—a golden retriever, a bottlenose dolphin, and an Asian elephant. These animals represent three distinct ways that we regard the creatures around us, views reflected by the titles of the three parts of this book. The dog, a creature intimately known to most of us, stars in Part 1, "Familiar," with another common animal, *Homo sapiens,* brought in for comparison along the way.

In Part 2, "Extraordinary," we broaden our view to include an animal that we've all seen, at least on TV, but that few of us get to know on a daily basis—the dolphin. They take a full year from conception to birth, six times as long as dogs. In this section, we take side trips to see what we can learn from the gestation of a red kangaroo and an emperor penguin. Every step of each animal's prenatal development reflects its history and predicts its future. Part 3, "Extreme," finds us moving even farther afield from our usual urban experience (at least in the West), with the largest animal on land, the elephant. Asian elephants take almost twice as long as dolphins to develop in the womb—22 months. To explore other extreme adaptations to reproductive needs, in Part 3 we check in several times with a lemon shark. We will even peek briefly into the monster-movie life of a parasitic wasp, at its horrific exploitation of other creatures to ensure the future of its own next generation.

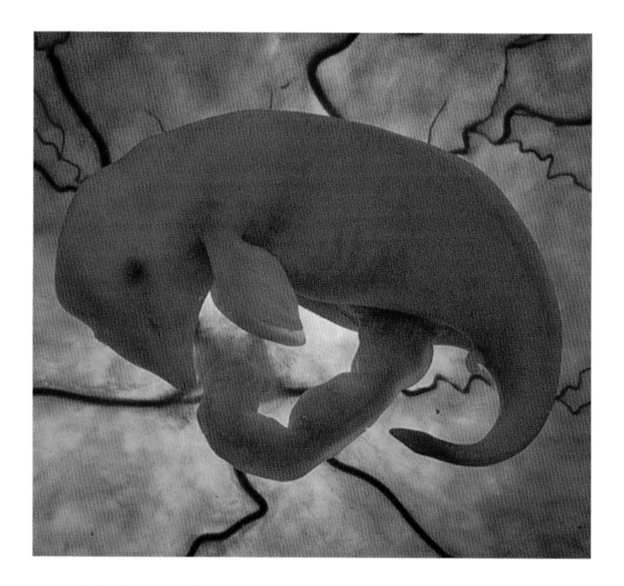

A dolphin embryo 6 to 8 weeks after conception

We begin each of our three main stories on the day that the future parents mate and set in motion the hidden artistry of the womb. Always preoccupied with making the next generation, nature has devised some eyebrow-raising mating strategies. Reproductive habits vary as widely among these animals as do their methods of development before birth, so we must glance at these before looking at their results.

Part of my fun in writing this book has been discovering what each animal's development tells us about its behavior, habitat, and evolution. You will find that history plays a crucial role in these stories—the histories of individuals, of species, and of whole classes and orders of animals. How did these creatures come to be shaped the way they are and behave the way they do? To what demands and opportunities did their ancestors respond? In which ways are they related? The answers may surprise you. We will discover, for example, that dolphins had terrestrial ancestors and that the forebears of elephants may have been aquatic.

Ring-tailed lemurs in Madagascar

The evolution of sexual reproduction is one of the great stories in the history of life on our planet. Unlike the solitary division and multiplication involved in asexual reproduction, procreation begins with addition—one plus one, mother plus father, egg plus sperm—to form a zygote that then divides and grows into a new individual. Biologists precisely define "procreation" as reproduction in which a male and a female produce offspring that differ from them. The offspring differ because each parent brings its own genetic heritage. We practice this method of babymaking ourselves, so perhaps we take it for granted. It seems "natural" to us. But consider what it makes possible: the union of two separate histories, a merger of strengths and weaknesses, of potential and risk. If you stop to think about it, it's a very strange method for manufacturing a new generation. Surely, with its endless time for research and development, nature could have come up with a simpler routine.

But it is precisely this vast amount of time available for experimentation, along with countless generations of eager

A peacock (a male peafowl) in full courtship display

experimenters, that has created the staggering diversity of life. Just think of the forms that living creatures can take—the Art Nouveau finery of a strutting peacock, ladybug beetles as detailed as jewelry, a baby's articulate hands. Deep-sea fish use light as bait and giraffes look like something that children would design. Crickets play themselves like animated musical instruments in Wonderland. Our world is not predictable.

We think of animal behaviorists as voyeurs who spy on zebras through binoculars, but we are all surrounded by animals. We can learn something about them from every move they make. Every individual is acting out its species' history and every new generation is potentially redirecting its future. Looking back into prehistory, we see animals evolving in the sea only to crawl onto land or developing on land and then heading into the ocean; dolphins made it all the way to aquatic while sea lions remain half at home on land. The very word evolve means "to unroll," and the evolution of a species unrolls slowly over vast amounts of time. Among the animals in this book, sexual reproduction makes this change possible. One-time cell mutations that occur randomly in these individuals would each be a dead end—could have no effect on the direction of their evolution—unless they were passed on to the next generation.

This is how animals adapt to their world. Every time they reproduce, these individuals slightly tweak the evolution of their own species. If you look into the past, you find that every animal carries within itself, as do human beings, a history that goes back to the beginning of life on Earth. But no one knows where this constant recombining of genetic heritage may lead in the future. So, in watching these animals as they develop in the womb, we aren't merely voyeuristically peeking at our cousins' private lives. We are watching the future unroll.

Giraffes and zebras in Africa

While planning this book, I happened to visit San Francisco. I had been on the phone with my editor, discussing how I would write about dolphins, only a few hours before my wife and I braved the tourist traps of Fisherman's Wharf to visit the famous sea lions that have colonized the docks of Pier 39 over the past couple of decades. With Russian Hill rising in the background, hundreds of sea lions laze across dock after dock—intimately huddled, barking constantly, and looking poorly upholstered except when emerging sleek from the water. California sea lions are pinnipeds. The term means "fin-footed," and their group includes seals and walruses. Because pinnipeds hunt in the sea but come ashore to mate and sleep and argue, biologists consider them semi-aquatic, in contrast to fully aquatic mammals such as the cetaceans—dolphins, porpoises, and whales. Although awkward out of the water, sea lions can propel themselves across ground with surprising speed. Recent molecular evidence indicates that they descended from a bearlike ancestor that lived on land.

A dolphin with hind limbs

I stood at Pier 39 thinking about the different ways that mammals have adapted to their environments around the world. Although still needing to breathe air, like every other mammal, dolphins eventually became fully aquatic. They seem as naturally marine as any animal might be; they even give birth while swimming.

The first time I saw a dolphin in the wild, I was canoeing the mouth of a river on the coast of South Carolina. Suddenly a circus troupe of dolphins was stitching the water of the estuary only a few yards beyond the bow, leaping into silhouettes against the sky and then diving again. I thought they were the most graceful creatures I had ever seen, their sleek form like some Platonic ideal of a water creature. So I was fascinated to learn about their prenatal incarnations. In the womb, dolphins go through a phase in which hind legs begin to develop and then disappear. This is further evidence that their ancestors were land animals. Sometimes this primordial trait re-expresses itself, just as now and then a human being is born with a tail or covered in hair. In 2006, Japanese fishermen caught a bottlenose dolphin whose hind limbs hadn't disappeared in the womb but were present and functioning as a second set of flippers.

Semi-aquatic California sea lions in San Francisco

ER 39
IER 39
PIER 39

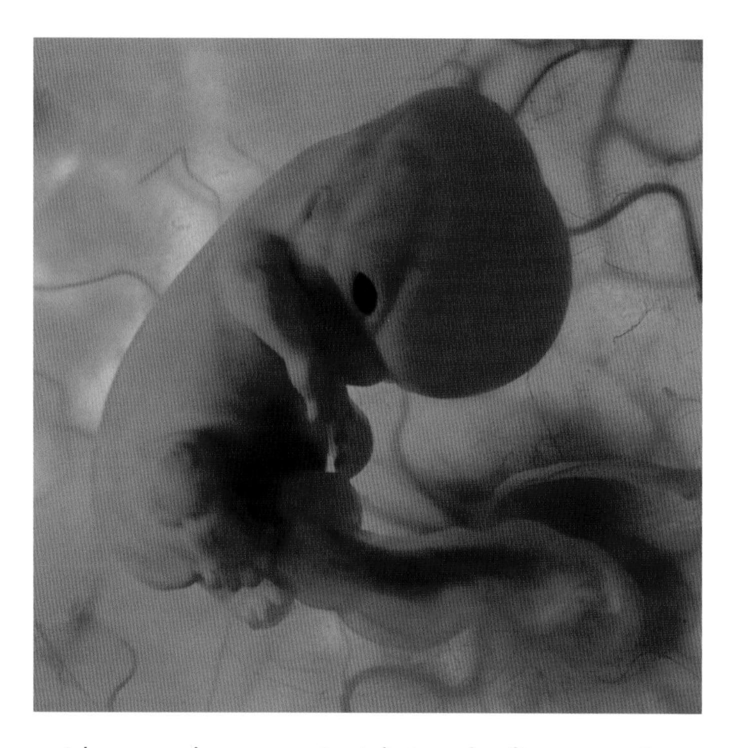

A human embryo approximately 8 weeks after conception

At Pier 39, crowds of humans stared at crowds of sea lions. I stepped back from the mass of tourists to photograph my clothed, upright fellow primates as they admired our distant kin. Two or three people in the crowd had dogs on leashes—a chihuahua, a fox terrier. Nearby a squirrel sat daintily nibbling the remains of a sandwich. Halfway down the boardwalk, a tortoiseshell cat navigated a fence railing as if walking a tightrope. Sea lion, dog, squirrel, cat, human—mammals all, wildly different, and yet with so much in common, like most of our characters in this book. At that moment I realized that this book would have to explore the commonalities and differences among mammals, as revealed by the way that our prenatal development reflects our unique history. Birds and sharks and other creatures show up, but most of our stars are mammals.

Our class, Mammalia, seems to be built of malleable clay. No other animal group exhibits such variety of forms. What a parade they have made, cavorting across the globe for millions of years: bats and armadillos, zebras and manatees, zoologists and acrobats. Scientists consider mammals to be keystone species in many habitats—those whose effects on their environment outweigh their population, whose loss might result in a tumble of ecological dominoes. Dolphins and elephants are keystone species in certain habitats, as are prairie dogs and sea otters in others.

Scientists speculate that mammals survived millions of years ago, long enough to diversify and flourish, largely because they evolved homeothermy—warm-bloodedness. The body possesses its own thermostat, resistant to the potentially damaging heat or cold of its surroundings, enabling mammals to colonize every habitat on Earth. Because each embryo and fetus builds toward its own future in the world, much prenatal growth is devoted to the young mammal's future need to maintain its temperature within a certain range. From the dolphin's blubber to the dog's pant, from the elephant's giant ears to the penguin's thick layers of interlocking feathers, each animal must develop in the womb the ability to stay warm in cold weather and cool in hot weather.

Another interesting point that emerges from these animals' stories is the relationship between development in the womb and growth after birth. Dogs need to nurse for several weeks because they are born with their body not quite finished. Kangaroos are able to give birth to far less developed young because they can then hide them in a pouch and feed them nutrient-rich milk until they mature. Sharks are born ready to fend for themselves. Elephants need years to grow up. Why the differences?

These points give you some idea of the fascinating questions we will explore in this book. I like to remind myself, however, that we are talking about individual animals. That fearsome creature "the shark" is as fictional as an "average teenager." In the real world there are only individual dogs and sharks and penguins, each as unique as yourself, each with a history that, like yours, goes back to the beginning of life.

We follow each animal from conception to birth, but there we leave them. Like other creatures, including human beings, these young animals must make their own way in a busy and uncertain world. They will face danger. Like us, they will not remember having already survived an adventure as old as their species yet always as new as their own body—the journey to birth.

Most mammals are born both unfinished and vulnerable.

Animal	Weeks 0–5	Weeks 6–17
Kangaroo AVERAGE LIFE SPAN 15-20 YEARS	0: female mates with more than one male 4: joey “born” and races to pouch to begin nursing; mother mates again, stores second zygote	10: joey has barely formed eyes but nurses constantly
Dog AVERAGE LIFE SPAN 10-12 YEARS	0: future parents mate in “tie stage” 5: embryo’s heartbeat can be detected	7: all fetuses touching; paw pads develop 9: litter of pups born blind, needing milk 14: pups are running and gamboling
Penguin AVERAGE LIFE SPAN 20-40 YEARS	0: lays huge egg 24 hours after mating 4: female goes to sea, male incubates egg on his feet	12: female returns; egg hatches
Human AVERAGE LIFE SPAN 75 YEARS	0: mate in any season in every climate 5: heart beats with reliable rhythm	12: all essential organs achieve basic form 16: hands begin to grasp and release
Shark AVERAGE LIFE SPAN 25 YEARS	0: female mates with numerous males	8: the embryo is the size of your little finger 12: iconic dorsal fin now prominent in womb
Dolphin AVERAGE LIFE SPAN 30 YEARS	0: hurriedly mate underwater, watching for predators	9: nostrils merge to form blowhole 11: flippers have formed 13: “melon” area (for sonar) grows in head
Elephant AVERAGE LIFE SPAN 65 YEARS	0: male and female mate without penetration 5: strong elephant heart begins beating	8: embryo resembles ancestor of all mammals 16: trunk is visible on embryo

Each line segment represents a different time scale.

Weeks 18–41

20 — oey still ursing, till hasn't pened yes

25 — joey releases teat and feeds only when hungry

30 — joey leaves pouch for first time

36 — joey leaves pouch for good

32 — some breeds reach sexual maturity

32 — first dive into the sea

20 — ead and dy covered ith hair

26 — survival outside womb possible on medical support

38 — born, often with difficulty because of large brain

28 — yolk sac has turned into a placenta

24 — fetus curled in U shape; muscular tail formed

28 — eyes opening and closing in womb

36 — now adult gray color and growing blubber

noving legs as if running, digestive system developing

Weeks 42–90

52 — joey no longer suckles

72 — joey goes off to fend for himself

72 — another generation could be born

52 — reach sexual maturity

52 — walking and talking possible

68 — a baby can find a toy after watching someone hide it

52 — pups born quickly and flee even their own mother

52 — born tail first; must rush to surface to breathe

52 — second kind of nose (vomeronasal organ) growing

76 — at 140 pounds, still gaining a pound per day

88 — calf born tail first like an aquatic animal

Jack Russell Terrier

English Springer Spaniel

Part 1 Familiar

Blue Great Dane

Partners

To most of us, dogs represent friendship—whether the fifth generation of prizewinning Irish setters in your family or the aging mutt rescued from the animal shelter and now dozing happily across your feet. Affectionate portraits of dogs crowd every rack of greeting cards; there are whole books devoted to *New Yorker* cartoons about them. We now recognize more than 400 breeds. Estimates of the worldwide population of domesticated dogs tops 400 million, and Americans alone spend billions of dollars annually on them. No wonder our scientific name for this species of the Canis genus is *familiaris*—familiar.

Dogs have been praised by everyone from St. Francis to Adolf Hitler. Around the world and throughout history, we have demonstrated their significance in our lives by their presence in our arts. In the half-millennium before the birth of Christ, craftsmen in what is now Germany created brightly colored glass figurines of dogs, complete with arched tail and perked ears. As recently as the 19th century, natives of the Solomon Islands, near New Guinea, were merging the facial features of humans and dogs in elegant wooden canoe prows and other sculptures, possibly because of their creation myth about a heroic dog who taught the first human beings how to live in the world. Dogs represent fidelity in 15th-century Dutch genre paintings and barely repressed wildness in the novels of Jack London.

Along with such creatures as cows, sheep, pigs, chickens, and horses, dogs have played an important role in our history. Some scientists theorize that without dogs, our ancestors might not have been more successful than the Neandertals with whom they shared the world for thousands of years. Since we first began to domesticate each other around cave fires, we have nudged the future evolution of dogs by carefully selecting for traits that would make them more useful—or at times simply more appealing—to us. Styles and sizes in the dogdom catalog range from the cartoon silhouette of the dachshund to the mastiff's intimidating bulk. Exploiting their territorial and herding instincts, we have bred dogs to guard our huts and superintend our flocks. Many breed names recall the jobs for which the dogs were originally bred: retriever and bulldog and shepherd; schnauzer for the smart nose, and terrier for dogs that go underground after other animals. "Pinscher" means "biter," and "dachshund" means "badger dog."

As we follow in the womb the fascinating growth of a litter of golden retriever pups, we will see how these impressive senses and abilities develop. This breed was officially established only in 1908, after decades of crossbreeding between the now extinct tweed water spaniel and the flat-coated retriever. Nowadays they are among the most popular breeds in the United States, with several established lines bred variously for show or hunting or service. The golden retriever's intelligence and obedient, affectionate nature make it as successful in guiding the blind as in tolerating the visiting grandchildren.

Like bonsai, dogs have been nudged into all sorts of shapes.

***previous pages*: Humans have long exploited dogs' pack instincts.**

Not Quite Ready Day 1: Part 1

No matter how much our careful breeding has flattened their schnoz or stretched their legs, canines from pug to borzoi go through similar patterns of mating and reproduction. Gestation usually requires about nine weeks (63 days). In the 22 months it will take our Asian elephant to have a single baby, this golden retriever can produce at least two generations. With as many as seven to nine pups per litter and each pup reaching sexual maturity at the age of nine months, in this time there can be a great many new dogs in the world.

The wild gray or timber wolf, *Canis lupus,* breeds only in the spring, but most domesticated female dogs usually enter a sexually receptive stage, called estrus or heat, about twice a year. Males, in contrast, are always looking for an opportunity to invest their genetic inheritance and will mate with any willing female. In packs of wild dogs, the dominant (alpha) male usually reserves receptive females for himself. Even without such haremkeeping, females much prefer acquaintances to strangers. This is one reason why breeders often have trouble coaxing a female to mate with a male that she met five minutes ago, no matter how impressive his pedigree.

Usually females prefer to begin the mating ritual with what looks to us like flirtation. A female may kneel in a position that biologists call the canid play bow. Waving her tail, she lowers her forelegs to the ground with her rear still up, as if stretching. She might use the same posture to coax a human being to take her for a walk. Coyotes, wolves, and golden retrievers all engage in this universal canine display. Dogs perform a free repertoire of ancient ritual all around us—the ears-flattened compliant tail wag, the pack animal's total submission of rolling over and revealing the vulnerable belly. When approached by a male, an unreceptive female may roll over into this submissive position. In general, mammals use scent in mating more than visual or auditory signals; a female that is ready to mate may sniff the male's penis and allow him to sniff or lick her vulva.

Just as hormones nudge the female to play flirtatiously, so do they trigger her ovaries to release eggs. Unlike many other species, which release mature eggs destined for quick degeneration if not fertilized immediately—a group that includes human beings—a female dog releases eggs that aren't yet ready to be fertilized. Working dogs and other large breeds release around ten or twelve eggs and give birth to an average of half a dozen pups; toy breeds release only about seven or eight eggs, usually giving birth to no more than one or two pups. When they begin their journey to the uterus, the eggs are still immature, but the female is ready to mate anyway. Biologists speculate that this slow maturation may be an evolutionary adaptation to the lifestyle of wild canines. If a female in a pack had difficulty winning a mate, delayed maturation could improve her odds of successful fertilization. As a consequence, individual pups within a single litter can have different fathers.

Immature eggs moving toward the uterus

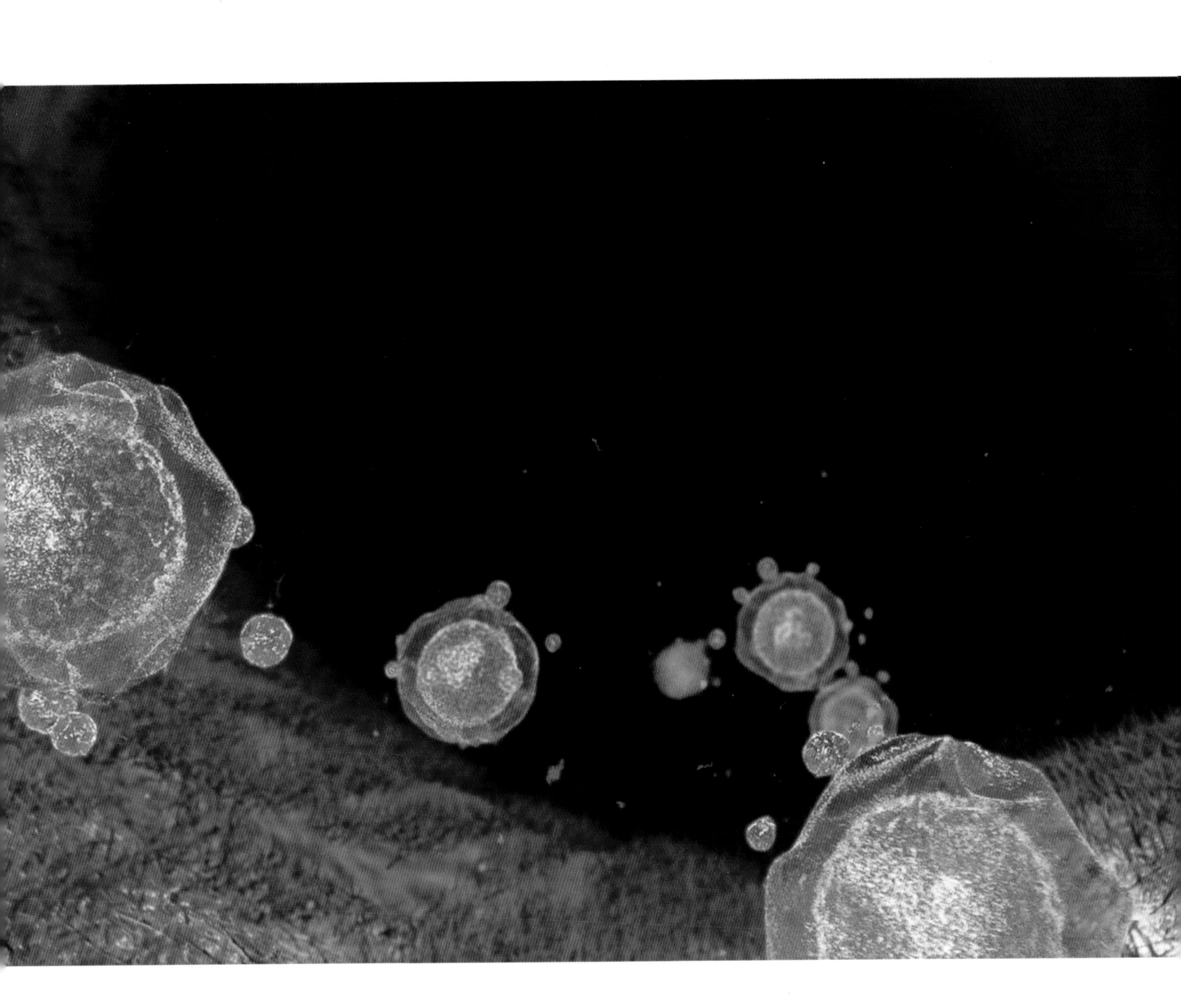

Self Storage Day 1: Part 2

Dogs are not the only animals that have evolved ways to extend fertility options. Numerous species of bats hoard sperm for up to several months. Some brown bats mate in autumn before going into their cold-weather hibernation. The females store the sperm as if in suspended animation. When spring arrives, the greater warmth and the longer periods of daylight stimulate ovulation, at which point the female bat's biochemistry finally introduces this year's eggs to last year's sperm.

Evolution came up with a different solution for hares, only 20 percent of which survive the first week following birth. In response to predators' war against them, hares have evolved large litters, frequent estrus, and their own unique twist on the reproductive cycle. In most species, hormones prevent females who are already pregnant from producing any more eggs until after they give birth to the current litter. The brown hare (*Lepus capensis*) has evolved a way around this barrier. After six weeks of pregnancy, a female can mate again and stimulate the release of new eggs from her ovaries, a process called superfetation. With 80 percent of her litter likely to die immediately, the mother can produce another litter in only a few weeks.

Like many other mammals, from rodent to primate—but excluding human beings—dogs have a bone inside the penis. There are a number of theories about why some animals have a penis bone and others don't, the most common being that this internal support evolved to help make mating more stable and productive. But those species lacking a penis bone, such as humans and horses, can also mate vigorously. The need for stability in dogs may result from the angle of penetration; the vagina is angled in such a way that the male's penis must enter it almost vertically.

Within seconds of mounting the female, the male dog ejaculates, although it takes another minute or two for the richer concentration of semen to emerge. Even then the two dogs don't separate. They can't. The female's vaginal muscles grip the rigid penis, on which erectile tissue enlarges during mating. Consequently the dogs lock together in the "tie" stage that provokes so much embarrassment when dogs mate in front of the children and neighbors. During the tie, the male's prostate gland releases more fluid to help transport the sperm to its goal. The tie maneuver has been observed in no other creatures besides canines. Normally the dogs spend this vulnerable phase, which can last anywhere from five minutes to an hour, back to back, failing to look nonchalant. Scientists theorize that this posture evolved to allow the participants to better defend themselves if a predator catches them in flagrante.

Because of the eggs' delayed maturation, when sperm reaches the uterus it attaches to the wall membrane and settles down to wait before continuing its swim. It can last a week or even longer but usually needs to park there for only two or three days. When the eggs reach maturity, a come-hither chemical signal tells the sperm to stop lingering in the foyer and proceed to the inner sanctum.

Many sperm reach the egg, but only one gets through.

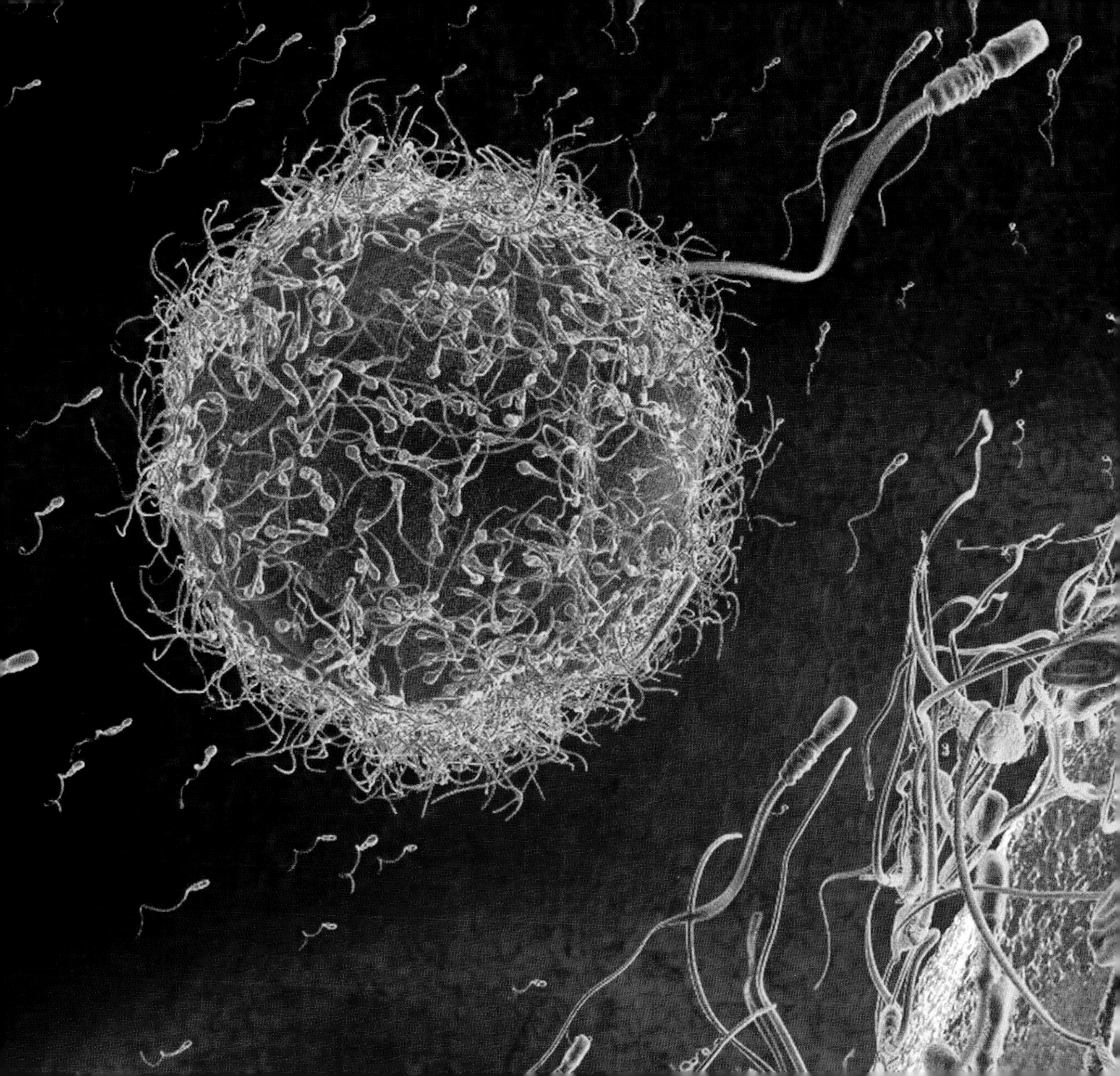

Evidence of Life Day 2 to Day 14

Biologists assign the term "growth" many definitions in a vast array of circumstances—development of body parts, accumulation of mass, replacement of spent skin cells. However, if we want to really start at the beginning for our golden retriever pups' journey to birth, we can say that their growth begins when the father's sperm fertilizes the mother's ovum, which soon launches the process of cell division. But the cells spend a long time growing before they can be called pups or even embryos.

When egg and sperm cell fuse, they merge into a single cell, a zygote. During a process called cleavage, the zygote begins multiplying by dividing—creating exact replicas of itself that separate into individual cells without otherwise growing. This new bundle of cells, the blastosphere or blastula, has two primary layers: an inner one that will form the embryo and an outer one that will form the placenta. So that our future pup can link up with its mother to receive its share of prenatal nutrients during the long growing process, the blastosphere must attach to the uterine wall, at which point it is called a blastocyst. (We will examine this attachment process shortly.)

It has been two weeks since our golden retriever mated, which means that she has already reached the one-quarter mark of her two-month pregnancy. The ball of cells has become more like a disk. At about this time in the development of any animal embryo, it enters a phase called gastrulation, during which it changes from a sphere of potential to an organized body-sculpting factory. Growth requires far more than merely an increase in size. Manufacturing anything as complicated as an animal requires a team of expert tissues, each with its own blueprint and schedule. There is a great deal of work to be done.

Gastrulation forms the germ layers—separate groups of cells that will eventually build different parts of the body. There are three primary layers, called ectoderm, endoderm, and mesoderm, for outer, inner, and middle skin; signs of them are visible even in the egg and zygote in their asymmetrical shape. The ectoderm constructs the skin and nails, as well as the retina and lens of the eye, and also creates the far-flung communication highway of the nervous system and its administrative capital, the brain. The endoderm forms glands such as the pancreas and liver, as well as the respiratory and digestive systems. The mesoderm builds the factories that eventually produce bones and muscles and another crucial aspect of our bodily infrastructure—a circulatory system to maintain the whole network.

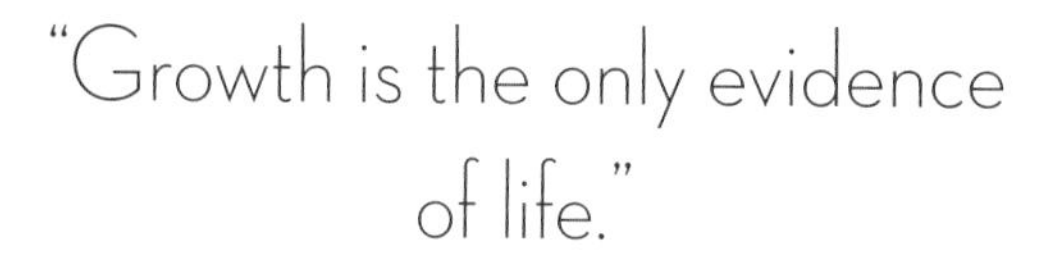

Cardinal John Henry Newman

In the 17th century, peering through his primitive microscope at a thin slice of cork, the English microscopist Robert Hooke named the structural compartment of living tissue a cell, the term for a monk's tiny room in a monastery. Hooke's image implies a stable and even rigid structure. Yet many cells are mobile, and sometimes their edges are more like national borders than like walls. Gastrulation restructures the fertilized egg into an embryo by directing cell migration. Long after the development of the embryo, throughout the dog's life after birth, other cells will need to migrate as part of its immune system or when rushing first aid to a wound. But by then these cells will already have been assigned their jobs. It is this first migration, the initial restructuring, that seems almost magical in the way it kneads new clay into material that will, over the next six weeks, sculpt itself into a dog. This process, which every living creature undergoes, is as mysterious and impressive as any discovered by science.

The molding begins when the primitive embryo folds in on itself to form a cylindrical tube. A longitudinal groove develops down the center of the embryo, a dividing line called the primitive streak. Gradually, cells move toward the groove and pour into it. The primitive streak establishes an axis around which the embryo can develop bilateral symmetry—the balanced design of two forepaws and two hindpaws, of matching ears and eyes and lungs. This kind of duplicate development, in which the right side mirrors the left, is common to all vertebrates. You can see it in a robin's wings or just by looking in a mirror.

Every cell in the embryo then flows over a region that instructs the cells to differentiate, as if they were receiving sealed instructions from the only officials high enough in the organization to assign vital missions. The German embryologist Hans Spemann, who was awarded a Nobel Prize in physiology or medicine in 1935 for his research into cell growth, coined the term the "organizer." It was also Spemann who first publicly explored the concept of cloning. He discovered that when he divided a two-celled salamander embryo into two individual cells, each grew into an adult, demonstrating that even the earliest embryonic cells contain crucial genetic information. The questions to be decided at this point in development are endless: Which cells go where? What has to be done first? How does a cell start growing and—just as important, when you think about it—how does it stop? The answers emerge not only without consciousness or planning but also without even a brain to respond to hereditary instinct. Such "decisions" are made on a cellular level, with biochemistry somehow directing itself.

Scientists theorize that it is during this process, called embryogenesis, that some kinds of cancer may result from a tumor in the germ cells. The possibility of a chemical misstep heightens the suspense as we peer deep inside the body of our golden retriever. But even when everything goes well—as it does most of the time, with other animals as with ourselves—the genesis of the embryo is an amazing and dramatic phase, the all-important first round of building an individual.

Next comes organogenesis, the construction of the various organs and their positioning in the ever changing body of the embryo. We'll follow these developments over the next few weeks. In only seven more weeks, these newly directed cells will have grown into a golden retriever pup.

A Big Family Day 15

Why do pups arrive in matching sets rather than individually packaged, as human beings usually are at birth? There are many reasons why the number varies in litters, broods, spawn, or whatever else we may call a group of siblings born or hatched at the same time. The number of dogs in a litter varies depending upon how we have modified the evolution of each breed, resulting in a wide range. The largest number recorded—far too many for the mother to care for efficiently—was 24, born to a Neapolitan mastiff in 2004. Incidentally, usually we reserve the word "litter" for mammals, whereas "spawn" describes the gelatinous masses of fish and amphibian eggs, and "clutch" designates the eggs of birds and reptiles.

Although twins are not uncommon, triplets and larger multiple births are unusual in *Homo sapiens* but the norm in many other species. It makes sense that profligate evolution would nudge animals to produce numerous offspring, considering how many are likely to fall to rapacious predators. Like Napoleon, nature can afford to lose many soldiers every day—but only because it keeps all its creatures frantically employed in perpetual reproduction. Insects, fish, reptiles, amphibians, and other animals lay eggs and fertilize them without parental investment beyond the relatively negligible cost of producing sperm and ova. For mammals, however, carrying offspring from conception to birth requires a huge physical investment from the female and often demands extended postnatal care from one parent or both. The same burden faces birds, which must feed and defend their nestlings. Therefore, numbers of young per litter vary greatly among species.

The time spent in the womb versus the time spent in caring for young varies as well. Not all species are born at comparable points in their development. As we will see with the growth of our three main stars, the young of dogs and dolphins and elephants come into the world very differently equipped, each with a different amount of growing left to do outside the womb. After the male and female golden retriever mate, nine eggs wind up fertilized—a bit above average for this breed. Having received their instructions from organizers, the nine tiny embryos disperse and attach to the uterine wall. This implantation initiates the development of an individual placenta that will nourish each embryo. The mother's body plants the embryos very precisely in her uterus, with the uterine muscles contracting to migrate them along until they are evenly spaced, with plenty of distance between them. At this sand-grain size, they seem to have been settled into vast estates, but during the next two months the embryos will grow until their borders touch and then crowd.

Implantation requires cooperation between the embryo and the wall of the uterus. First, microvilli—microscopic protrusions from the cellular membranes of both uterus and embryo—must intertwine like the grasping parts of Velcro. Then, securely anchored and evenly spaced, the embryos can settle into the business of growing and developing into an entire litter of pups. The two categories of human twins tell us a lot about how multiple births work. Fraternal twins, the more common type,

The ball of cells implanted in the uterine wall

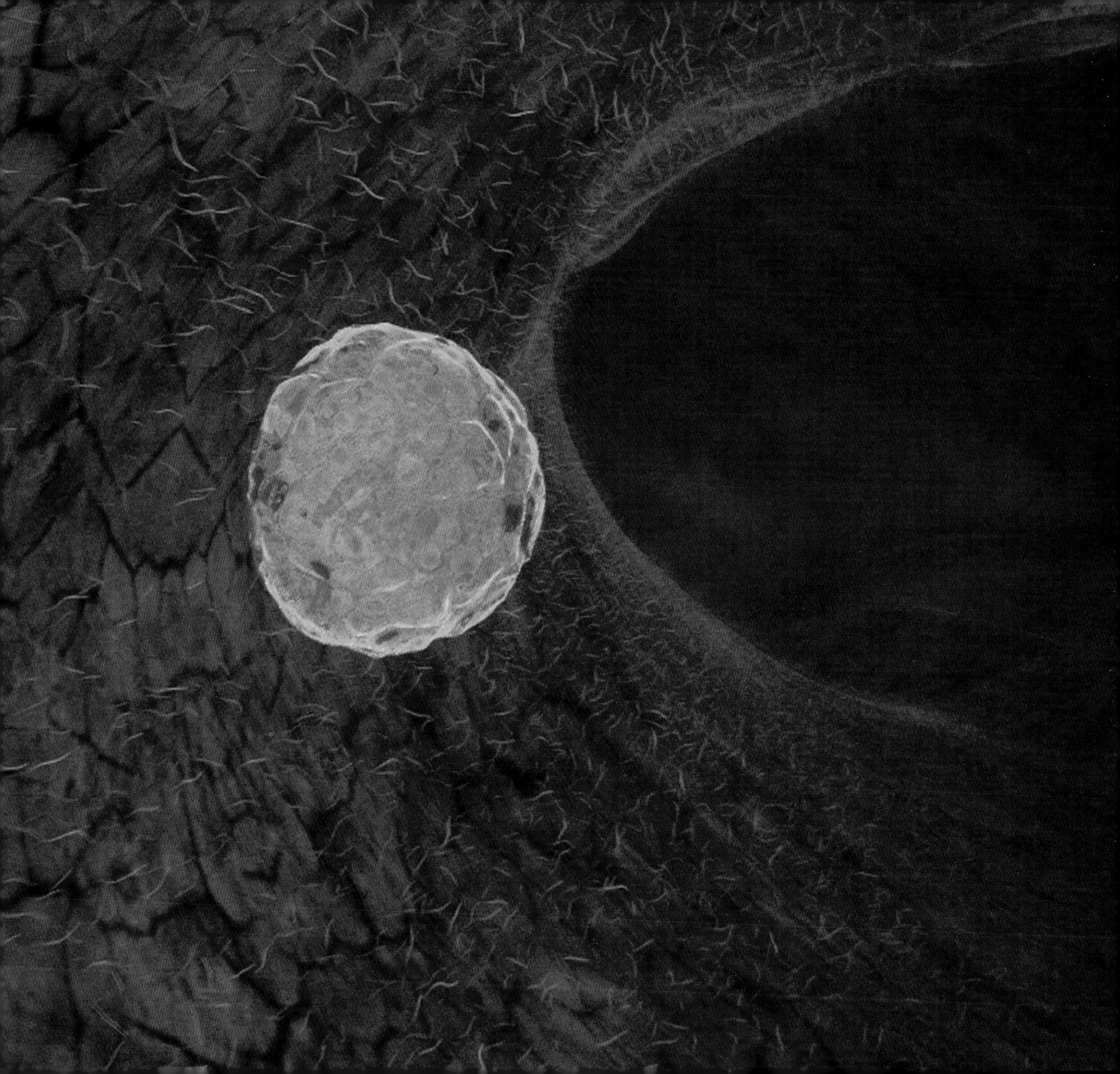

are dizygotic, growing from two separate zygotes. Two individual sperm fertilize two separate eggs, so that the resulting siblings are no more identical than any other siblings. They may or may not be the same gender. They share parents, but the different hand of cards dealt to each sperm and egg results in a slightly different way of playing the game. Identical twins, in contrast, inherit precisely matching genetic histories because they are monozygotic; they grow from a single fertilized egg that divides into two embryos early in the development process. The only way they can acquire any genetic difference is if a chromosome in the mother's mitochondrial DNA, which exists in the cell but outside the duplicating nucleus, mutates after fertilization. Such mutation, the very medium of evolutionary change, incorporates a tiny misprint into the genetic instruction book–a change that might modify the species for generations to come.

Identical twins are touching by the end of the first trimester and soon press against each other along bodies, limbs, and even mouths. As they each grow to know their own body, clenching fists and stretching toes and yawning, they become aware of each other. Scientists think that this close physical connection during prenatal development helps explain the strong bond that twins feel in childhood and even as adults. Throughout the development of fraternal twins, though, a membrane separates them. They never actually touch. In a similar manner, each golden retriever embryo develops inside its own membrane, the sac that will still surround it when it leaves its mother's womb to face the world. This membrane is not the same thing as the placenta itself, which will serve as liaison between mother and pup.

As we observe the growth and development of these embryos into fetuses, we find ourselves watching for the moment when they look recognizably canine. Well before birth, most mammals clearly resemble their parents. Therefore it's worth remembering that, although scientists' estimates of the number of animal species on Earth range from 3 million to 100 million, fewer than 4,400 of those (so far) are mammals. A great many young animals in the world don't look at all like their mother or father. Without prior knowledge, who could observe a translucent comma-shaped tadpole and a squat bullfrog in a pond and imagine that one is the offspring of the other? The lime green, wormlike caterpillar of the eastern tiger swallowtail looks nothing like its dramatically patterned yellow-and-black parent. Even birds take time to grow into familial similarity; unlike their tall, tuxedoed parents, newborn penguins are fat, round, and fuzzy.

The similarity between young and adult in dogs is one reason the species appeals to us so strongly. There is an evolutionary history behind why we find pups and kittens so irresistible. Their young go through changes similar to those our own children embody. Born with oversize facial features, just like baby humans, they trigger nurturing responses in their two-legged companions.

Only mammals have the ability to produce food for their young.

Phantom

By the third week of pregnancy, the embryos are still only about the size of a pea, but the golden retriever's pregnancy is beginning to be visible to her owners. As many pet owners have discovered, however, a dog doesn't actually have to be pregnant to exhibit signs that look like pregnancy. Nor are we misinterpreting her cues. All female dogs go through a pseudo or phantom pregnancy that mimics all the outward signs of a real one. You may have seen your dog carry toys to her bed and settle them around her, work her bedding into a cozy nest, even defend her imaginary puppy-raising territory. Some try to nurse the pups of another female.

Scientists think that this curious behavior dates back to dogs' ancestral past, well before they started cohabiting with humans. These retrievers in the womb, like every other domestic breed on Earth, are close cousins of the timber or gray wolf, which seems to have emerged as a distinct species roughly 300,000 years ago, in the late Pleistocene. In the wild, breeding is the perquisite of the alpha pair, nature's insurance that the species' strongest members thrive and carry a hardy bloodline into the future.

But this situation creates some problems. The dominant female is not only the mother for the entire pack, almost like a queen bee, but she's also the group's primary hunter. If she had to stop hunting while rearing her own pups, the pack would lose her crucial skills as breadwinner. Even for a smart alpha wolf, to go out hunting heavily armed elk every day was to court death, which would have left a litter of motherless pups back home. In an impressive adaptation, evolution contrived a way to avoid this potential shutdown of supply lines by having all the females in the pack enter the breeding season at the same time.

The evolution of the phantom pregnancy helps ensure the survival of both pack and species. Not only do all the females behave as if pregnant; they are actually able to produce milk. After the alpha female gives birth, she can return immediately to hunting because the unmated females are ready to take turns as wet nurse for her pups (see page 66). Even if one or more of the pack's females die—even if the pack loses the dominant female—the pups will still survive and grow up to reproduce, playing their role in the ongoing saga of mating and reproduction that seems to be the only thing nature cares about.

Breeders must determine whether a female is pregnant or experiencing a phantom pregnancy. For other species, a veterinarian could perform a blood or urine analysis for the presence of the hormone progesterone, but after female dogs ovulate, they always have progesterone present in bodily fluids. The best way to find out if your dog's pregnancy is real or phantom is to have a vet examine her, possibly with the use of ultrasound imaging.

Phantom pregnancy inspires many kinds of nesting behavior.

Seeing Things Differently Day 30

In the last ten days, the embryos have grown only from the size of a pea to the size of a grape. Yet within this tiny bundle of life, a half-formed heart has already begun to beat. A pup's heartbeat may be as rapid as 200 beats per minute, with an adult's average ranging between 50 beats per minute in larger breeds and up to 160 in smaller ones.

Until this point, the dog embryos have been developing at about the same rate as that of a human being. Now, however, halfway through their life in the womb, the future pups begin growing at a much faster rate. They don't have two-thirds of a year in which to lounge around in the womb before birth, as a human being would at this stage. A month from now, they must emerge into the world and begin life as dogs.

Reproduction in litters means that no dog embryo ever has a room of its own. At this stage, we can observe the strange way in which the nine embryos respond to their claustrophobic nursery. At the point on the embryo where the umbilical cord attaches, there is a curious bulge. It holds the intestines—on the wrong side of the abdomen. For a while the intestines grow outside the body cavity, within the umbilical cord area, because there is no room for them indoors. Their complex tissues need time to develop before birth, but there isn't yet space to store them inside the barely formed embryo. Only later do they travel inside and settle into place before birth.

The womb is becoming very busy. By this time, the embryo also has well-formed eyes. An early version of the cornea, the exposed outer layer of the eyeball, is in place. This remarkable feature of the eye must be transparent to admit light. The sclera, the white of the eye, is composed of cartilaginous or fibrous tissue that covers every part of the eye except the cornea. But the need for transparency means that the cornea, although it is an extremely sensitive area supplied with more nerve endings than any other part of the body, can't be nourished by the usual system of red blood cells that maintains everything else. Consequently, it gets its oxygen not from blood but from tears.

A few days ago the eyelids began to appear as folds of skin above and below the eye; gradually they grew toward each other and met and fused. Out in the world, one of their jobs will be to spread the oxygen-rich water from tear ducts across the entire surface of the cornea like a windshield wiper. Dry eye, inflammation resulting from a reduced tear film and leading to a cornea deprived of nutrients, is a common problem in dogs—and in human beings who wear contact lenses or spend too much time in air-conditioned rooms.

For now, however, the eyelids produce a substance that seals them closed, to protect the cornea from the unfriendly waste products in amniotic fluid. Although the placenta and umbilical cord filter out most waste produced by the fetus, its kidneys are polluting the uterine environment with urine that leaks into the amniotic fluid. Until after birth, the eyes remain sealed against possible contamination. This quarantine of sensitive organs demonstrates yet another way that evolution sometimes tweaks a process to get around difficulties caused by a malfunction elsewhere in the system. Evolution, as British scientist Richard Dawkins has remarked, "never starts from a clean drawing board. It has to start from what is already there."

Light enters the front of the eye through the cornea; the lens then focuses it onto the retina at the back of the eye. Beyond this point common to all mammals, canine eyes differ from ours in interesting ways that explain some of dogs' behavior. The human retina has two kinds of light-sensitive cells dubbed, because of their shape, rods and cones. We need both. Rod cells are more numerous and more sensitive to movement, but they ignore color. Cones respond to color and also provide greater visual acuity, like a digital camera with higher resolution. Sensitive to changes in intensity of light, they automatically adjust to differences such as that between looking out the window of a car and looking down at its speedometer.

Color-blind people and animals often perceive sheen and pattern better than creatures who see color. A camouflage pattern in dappled light, for example, is hidden from color-sensitive eyes but reveals itself to the color-blind by its luster.

Dogs have a different percentage of rods and cones in the eye than we do. Also, in contrast to our three kinds of color receptors in the retina, they have only two, which results in something like our color blindness. Contrary to what scientists once thought, the world of dogs isn't quite black and white. They can perceive the groups of yellow to green and red to orange and blue to purple, but they can't differentiate between members of each group. Thanks to the differences between their eyes and ours, they can detect a distant squirrel's movements before you can, but sometimes they have trouble locating a neon yellow tennis ball that you just threw into green grass not 20 feet away. Yelling "It's over there, you moron!" will not improve their color perception.

Scientists think they understand the ancient hereditary reasons for the difference between color vision in humans and dogs. Evolving in the shadow of giant lizards that prowled during the day, most carnivorous early mammals seem to have been nocturnal; there were more job opportunities on the night shift. Sensitivity to light was of greater importance to these mammals than sensitivity to color. Much later, our own omnivorous primate ancestors dined on fruits and berries that can be found only when their reds and yellows stand out from the surrounding green foliage. Scientists think that our color perception—that luscious sense that brings us the pleasure of both real peaches and Cézanne's—may trace back partly to this primordial diet.

following pages: **At 30 days, the paws and tail look canine.**

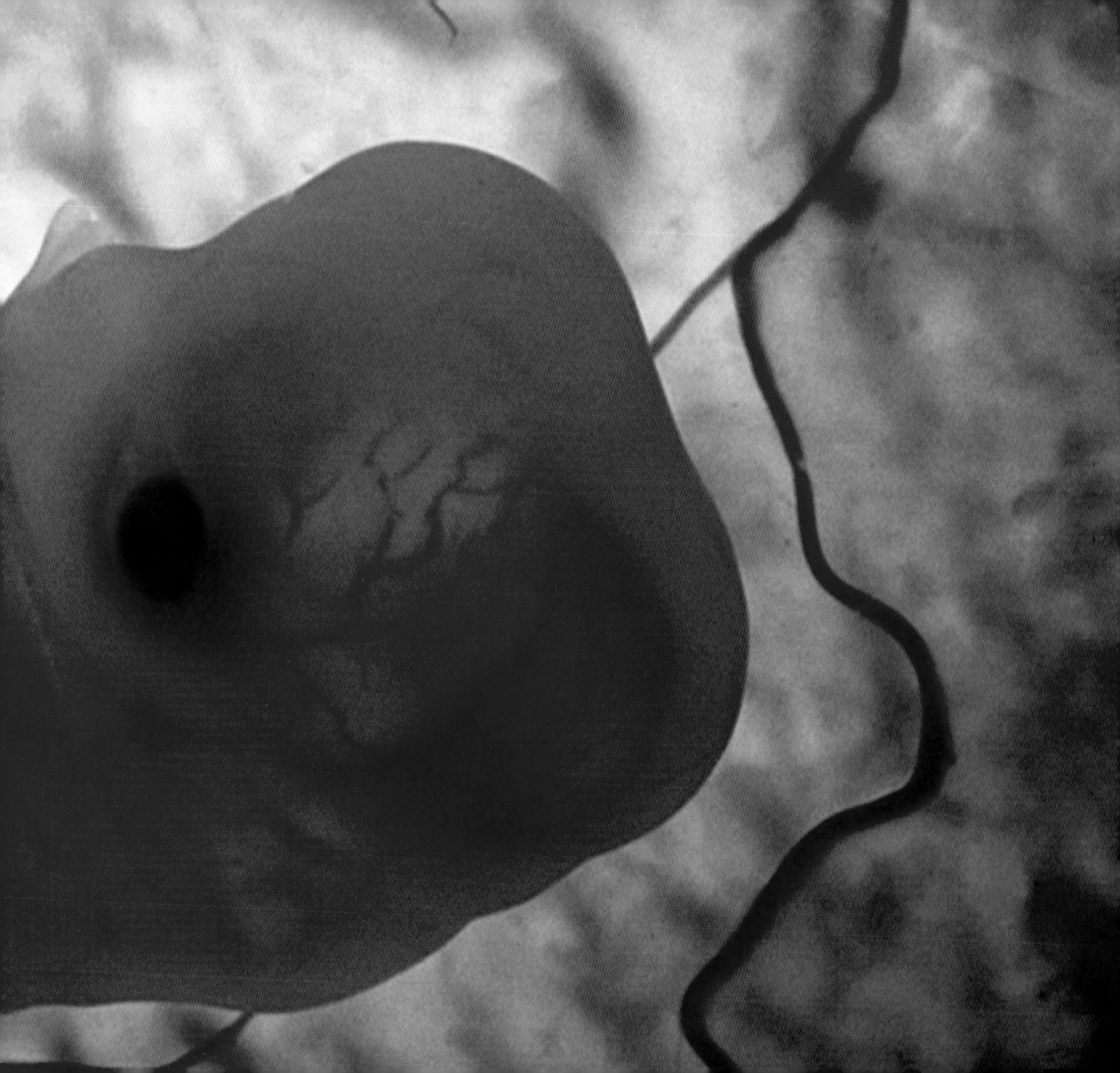

Running in Place Day 33

Although many dogs nowadays are confined to urban apartments without enough exercise, the natural state of canines seems to be one of raucous abandon. As rowdy pack animals that were later bred to hunt and track and herd, they naturally cavort. They pursue; they investigate; they warn and defend. Their legs evolved as efficient tools for many tasks besides trotting between their food bowl and their bed near your chair. Every time your sleeping dog's ears perk up or it turns around a few times before lying down, every time you take a pup to obedience school and it submits to your alpha dominance, you glimpse the ancient canine heritage.

The retriever's womb is building into her nine future pups all the exquisite senses that reflect their group lineage and predict their individual future. Now, a little more than halfway to birth, the embryo has all its organs in place and we can recognize its shape and structure as that of a dog. At this point scientists promote the embryo, which is now the size of billiard ball, to a fetus. Informally we might say that, in contrast to the obviously canine fetus, the embryo was more like an animal template, at first not even distinctly mammalian (see page 122).

The legs are now forming—perhaps the most dramatic of many changes taking place. Limb buds first appeared more than a week ago, and already the paws are distinct, paddle shaped and webbed with ridges. The major muscles are well developed and aren't waiting for birth to start moving. Like a human or elephant fetus (see page 142), a fetal dog engages in warm-up exercises long before it leaves the womb. Early movements are spasmodic, but fetal movements gradually develop coordination. This early motion prevents the muscles' surfaces from growing together and locking joints.

No canine places among the fastest animals on land. A cat takes the prize—a cheetah, dashing 70 miles an hour. Surprisingly, many predatory mammals are left behind by a racing ostrich, which can manage 43 miles an hour. In the water, the winner is the sailfish, which can match a cheetah's speed. A greyhound, because its supermodel physique comprises mostly muscle, can run 43 miles an hour.

Dogs' front legs carry the majority of their weight, 60 percent or more, while their hind legs are built more for going from zero to racing speed with one jump-start. In the womb, the fetus is now growing three kinds of muscle. The cardiac muscle is most specialized. A second kind controls movement of the organs—blinking the eyes and twitching the ears, keeping respiration in its essential rhythm. These perform their jobs automatically, without the dog's attention. The only muscles that the dog can bring under its control are the skeletal muscles, which lift a leg near a bush or turn the head toward a whistle or snatch a hotdog from a picnic table.

The distinctive paw pads are already forming.

A Larger Piano

By day 39, the inner ear is completely formed. This labyrinth inside the ear cavity, in the wall of the skull, has several jobs in the adult dog, as it does in human beings. Its fluid-filled caverns and semicircular canals serve as a balance system, determining position in respect to gravity whenever the dog tilts its head or accelerates. This region enables it to walk, chase prey, and right itself after a tumble. But these gyroscopic balancing maneuvers are so hidden and so automatic that we don't think much about how they work or from which hidden control center they operate.

What we usually notice about dogs' ears is much more obvious—how incredibly sensitive they are to every noise around them. The floppy cartoon ears of a basset hound, the elegant wolf ears of a Siberian husky—even in sleep they shift toward noises, monitoring the world from a distance. When you speak to your dog, it pricks up its ears in your direction and the outer ear bounces sound waves deep into the inner canals. The sound-gathering machinery of the external ear requires 17 muscles for all this perking and turning, raising and lowering.

Our own seashell-like outer ears gather sounds in the manner of a satellite dish, but we perceive nothing like the range of sound that a dog can hear. Their ears leave ours looking rather primitive. On average, your dog can hear a sound four times as far away as you can and, tilting both ears toward it, is able to locate the source within .06 seconds. For comparison with ourselves, let's translate canine ears' sensitivity into quantitative terms. Sound is a traveling wave of oscillating pressure. The unit of measure of sound waves, in cycles per second, is a hertz.

A hertz is actually the standard measurement of frequency of any kind, so if you wanted to, you could measure your heartbeat or respiration in hertz or one of its multiples. Usually, however, physicists employ hertz to measure sound waves and other forms of electromagnetic radiation.

Different wavelengths of sound tap the eardrum with different amounts of pressure, which we perceive as variations in pitch. The shorter the wavelength, the higher the pitch. Many sounds are ultrasonic, too high for us to be aware of at all; bats and other creatures communicate in this range. Because individual sound waves are so small, scientists measure pitch in kilohertz (abbreviated kHz), or 1,000 hertz. Human beings can register sounds up to 20 kHz—20,000 vibrations a second. A cat tops the record with 25 kHz, but a dog can detect 35 kHz, or 35,000 vibrations a second. How high is this in everyday terms? "To produce the top note a person can hear," says Stephen Budiansky in *The Truth About Dogs,* "would require adding about twenty-eight extra keys at the right-hand end of the standard piano keyboard. To get to the top dog note would take forty-eight keys, or four full octaves." No wonder their ears never stop twitching (see pages 94 and 116).

Dogs evolved as predators that needed to hear the high-pitched calls of rodents—a curious thing to remember when your dog is chewing a squeaky toy. He may live on kibble and snacks, but his inherited hearing system can still detect both the squeaks of mice and the growl of a potential rival a block away.

With its attentive ears, a wolf never stops monitoring its surroundings.

DVDs and Marijuana Day 40

By day 40, with only three weeks left to go until she gives birth, the golden retriever is looking visibly pregnant. Larger breeds take longer to show the telltale bulge, and sometimes it appears suddenly, as if overnight. Other muscles besides the legs are practicing the movements that they will perform out in the world. For example, ultrasound films reveal that even in the womb, well before birth, dog fetuses go through all the motions of panting.

An animal body is an ecosystem; it requires constant monitoring by its internal network to maintain an overall equilibrium. Warm-blooded animals have evolved a number of ways to direct heat out of the body to avoid overheating and damaging the system. Humans perspire through glands scattered over much of the body. Canine sweat glands, however, are limited to the pads of the paws. Instead of sweating, dogs release heat by panting—opening the mouth and exposing the tongue to evaporate water through it. An overheated dog that moistens its tongue again and again as it pants is doing so to speed up the evaporation that is lowering its body temperature (see pages 88, 100, and 146).

Of course, much of this same equipment will also be employed to produce a bark, that impressive tool that dogs use to greet, notify, and threaten, as well as to torment writers who are trying to concentrate. Wolves howl and coyotes yodel, but dogs bark, whine, yip, screech, yap, whimper, yelp, and wail. All sorts of historical records indicate that from the earliest days of our partnership with dogs, we were breeding them to enhance their barking voice. Back then, dogs were hunting partners and security guards; we needed vocal sidekicks.

The fetus's entire respiratory system is developing quickly now. Although, as we have discovered, dogs possess efficient ears and eyes as well, they depend primarily on the nose. Their nasal cavity is much more elaborate than ours, both larger and better informed, its wet nooks and crannies densely populated with sensory nerves. These reside in the olfactory version of the epithelium, a shallow layer of cells that lines most ducts, organs, and internal cavities, as well as covering exposed surfaces of the body.

How much more sensitive is this advanced olfactory system than our own? Just look at the numbers. Human beings have "only" about five million scent receptors in the folds of their nasal passages. Dogs, in contrast, have more than 200 million. Consider the impressive range of scents that most of us can distinguish: lemonade and rosemary; the differing amounts of cacao in dark and milk chocolate; pencil lead, hot pavement, old photographs, rubber, salt. Now imagine this variety greatly multiplied as well as amplified in intensity—as if the volume had been turned up—and thereby made to reach farther and last longer. Studies indicate that dogs can detect some organic chemicals at one percent of the concentration required for humans to smell them. In some experiments, scientists lightly fingerprinted a piece of glass and left it outdoors to weather for two weeks—only to find that dogs were still able to detect a scent.

The dog also boasts something called a vomeronasal organ or Jacobson's organ, located at the base of the nasal cavity but operating separately from the other scent receptors. Many animals, including human beings, possess this specialized odor

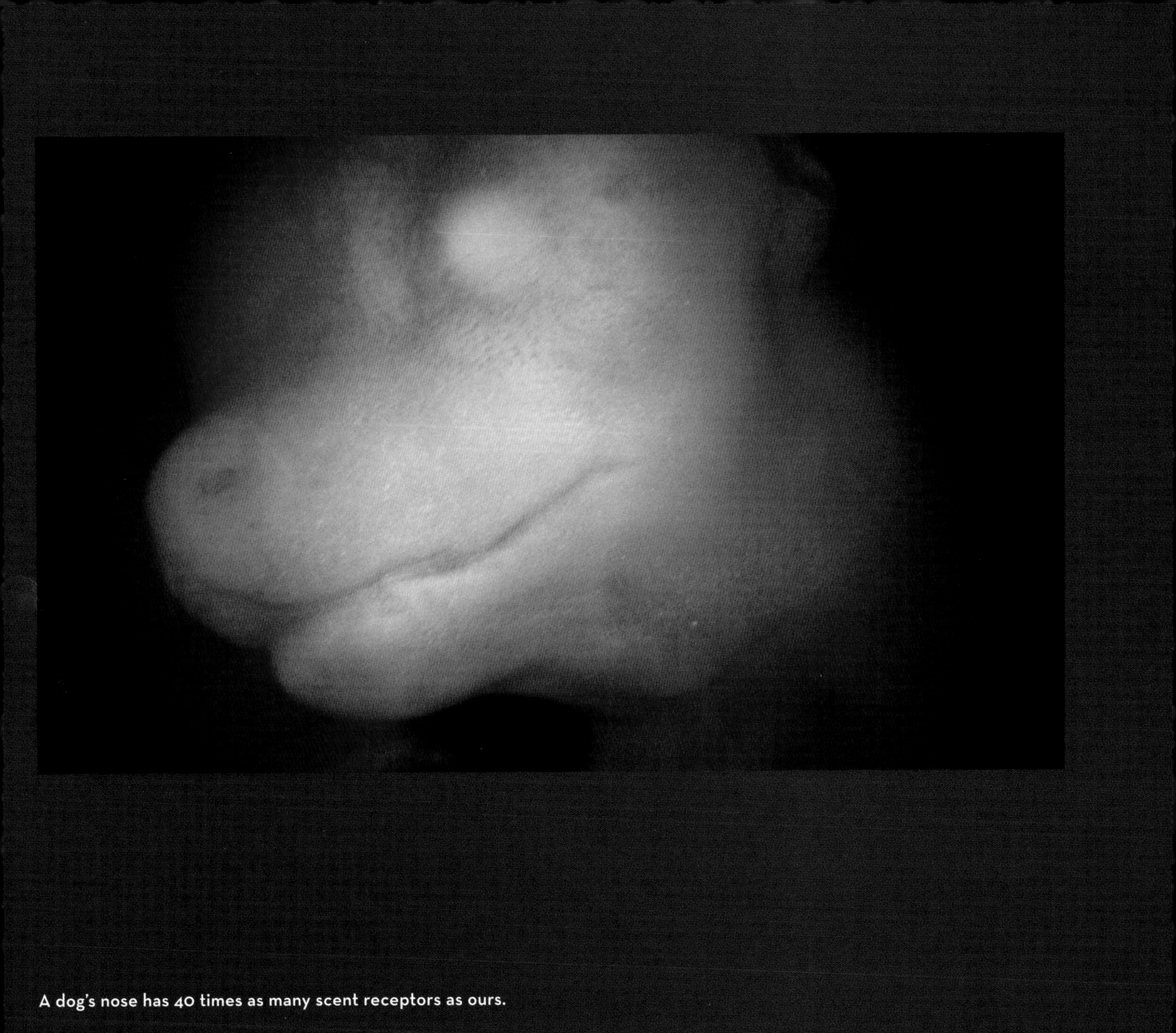

A dog's nose has 40 times as many scent receptors as ours.

detector, but its degree of importance varies, and scientists don't yet fully understand its complete role and influence. It is known, however, to participate in detecting pheromones, chemical signals between animals of the same species, and in processing them differently from other odors (see page 142).

Because these quickly growing fetuses will become golden retriever pups, they may wind up using their powerful sense of smell to assist human beings. Smart and obedient, retrievers are popular as crime detection dogs. But there are job opportunities nowadays for many species of canine in police and military work. We may have fewer shepherds and cart dogs than in the past—less need for a dog to help bring home dinner—but we now have lots of other ways of working with them. They assist the blind and deaf, pull wheelchairs, and use their specialized nose to help us in many other ways.

Many of us have undergone the disconcerting experience of having our luggage sniffed by a dog in an airport or finding our car investigated by one at a border crossing. Dogs are employed to detect blood traces, drugs, and weapons and to find lost children, disaster victims, and decomposing corpses. They can even be taught to sniff for polycarbonate odors to detect bootleg electronic items such as smuggled DVDs. In Australia in 2002, a prison sniffer dog detected a woman's stash of marijuana as she attempted to smuggle it into a Brisbane prison. She had hidden the marijuana in a balloon that she then coated with pepper, coffee, and petroleum jelly, after which she hid it in her bra. The prison's marijuana-focused dog found it anyway.

But even these illustrations don't quite equal the most impressive demonstration of the future sensitivity of the olfactory system growing in each of these fetuses. A good example appeared in a paper written by several scientists and published in a 2006 issue of the journal *Integrative Cancer Therapies*. The research team reported that dogs that had not previously been trained in any kind of forensic work were taught to sniff the breath of various people and distinguish between patients who had lung cancer and controls who did not. The dogs were able to detect incredibly subtle biochemical markings in human breath—clues that could not be detected by medical tests. For decades some scientists have been claiming diagnostic uses for the canine nose.

If other scientists can replicate these results, dogs will have raised the bar again in the definition of man's best friend. However, these developing retriever pups will need their most specialized sense long before they reach adulthood. Born with their eyes sealed shut, they will use the already functioning nose to find their mother's teats and begin to nurse.

Lucky and Flo, the black Labrador retrievers who became famous in Malaysia for locating pirated copies of movies, earned a scary tribute: Crime bosses put a price on their heads. Police had to move the dogs to a secret location, as if they were in a witness protection program.

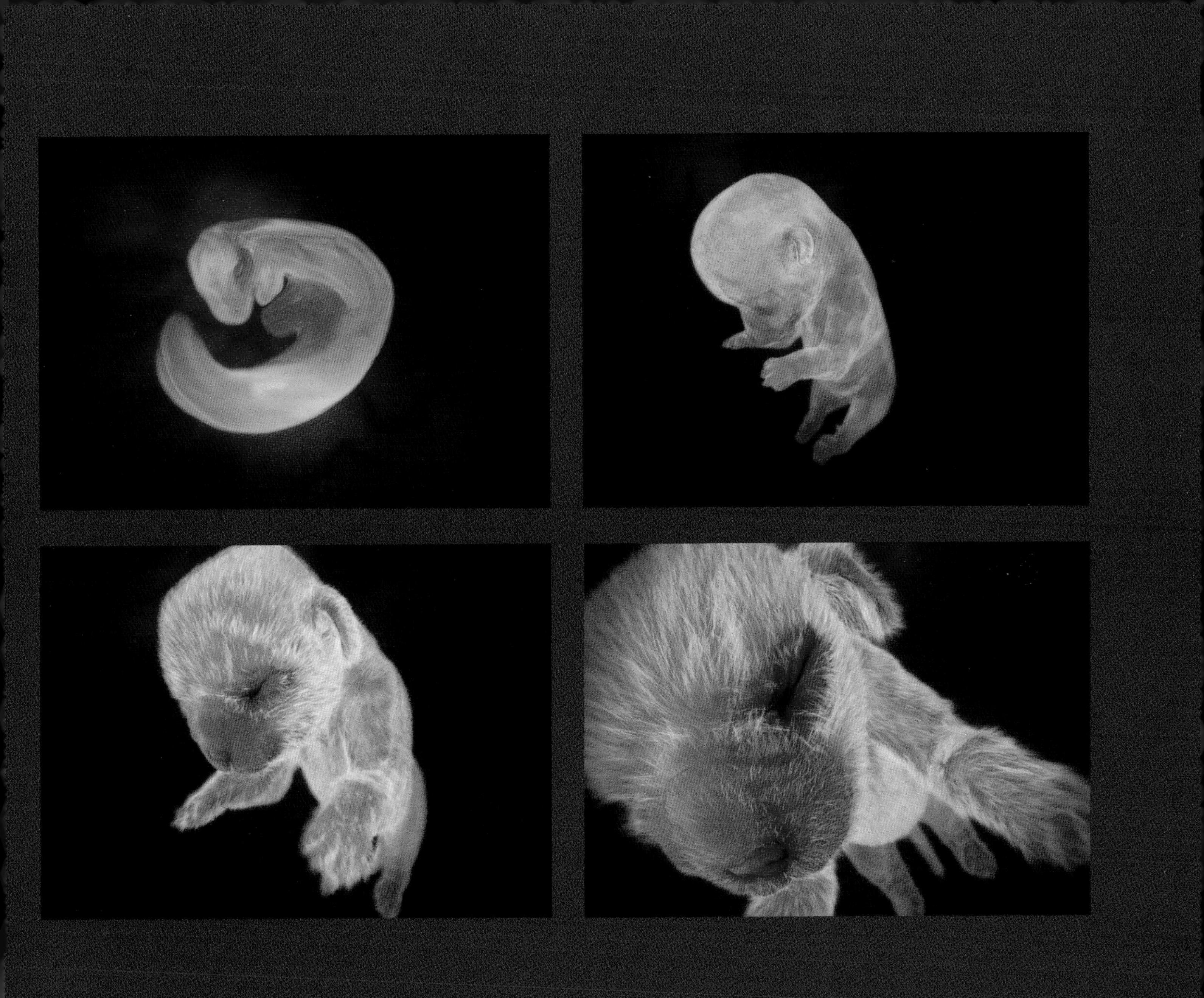

The dog's sensitive nasal system develops early in the womb.

Standing on Tiptoe Day 53

Halfway through the seventh week of their growth in the womb, the nine fetuses have now grown so large they touch each other. Crowded into this small space, going through panting and running motions, they create a rippling movement visible on the outside of the mother's abdomen. The formerly naked pups are now all handsomely dressed in fur that is a light cream color. Not until the weeks after birth will the pups grow into the rich hue that gives golden retrievers their name.

The pups' feet have also developed fleshy paw pads. Like us, dogs walk on cushiony callused feet. This is the thickest, toughest skin on the canine body, containing an insulating layer that makes it much less sensitive than the rest of their body to heat and cold. It is also the only part of the body through which a dog can sweat, which normally has the efficient side effect of keeping the pads supple—despite the daunting range of environments they face in the course of a day.

You may have noticed that dogs run around on tiptoe, like a woman in high heels. Dogs are digitigrade; they walk only on the tips of their digits, like deer and cats. We, in contrast, are plantigrade, walking on the soles of our feet with the heel touching the ground, like bears. In dogs, the part that would be the sole of our foot or the palm of our hand is elevated, tilted so that the heel doesn't touch the ground. Digitigrades tend to move more quietly and quickly than plantigrades and other kinds of animals. As we will discover when we follow its development in the womb, the Asian elephant is actually semi-digitigrade, but the unusual weight-bearing structure of its foot makes it appear plantigrade.

Naturally the dog's jacked-up posture has an evolutionary story behind it. Take a moment to examine your dog's feet. Only four toes touch the ground, each equipped with a claw that doubles as defense weapon and digging tool—and that has the useful ability to renew itself. On the front paws, too high ever to touch the ground, you will see the fifth toe's claw, called the dewclaw. As the dog's predatory ancestor evolved to chase prey, its legs grew longer and narrower, and the dewclaw receded to a vestigial position.

Like your own fingernails and eyebrows, like the horns of the narwhal and the rhinoceros, like the hooves of a Shetland pony and the feathers of a bluejay, dogs' claws are largely composed of the durable protein keratin. Almost insoluble, keratin is so strong that it inspired the perennial myth that hair and nails grow after death. Actually, as the tissues and skin decompose and shrink, hair and nails merely become more prominent and outlast everything else except bones. Claws and nails are produced by the skin as an extension of it, rather in the way that the skin of a shark produces its teeth (see page 138).

Now a useless remnant, the dewclaw recalls forgotten ancestors.

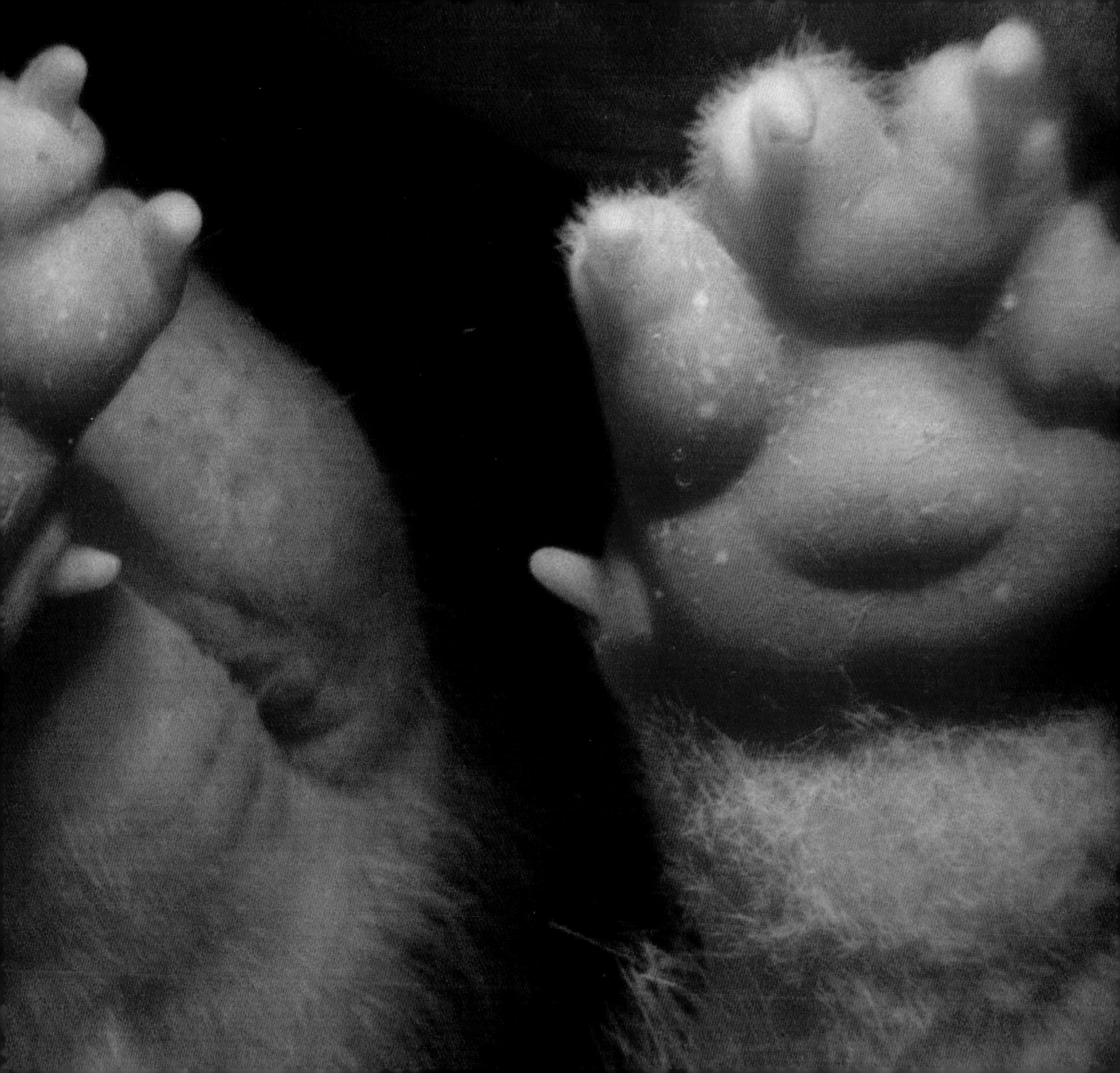

Bewhiskered Day 55

Besides the fur upholstering their fat, round bodies, these fetuses have also produced some other remarkable hairs. Around the 30th day of pregnancy, the embryo began to develop a number of small dimples near the mouth. Now, with barely more than a week left before birth, you can clearly see where the dimples turned out to be follicles that have produced dozens of well-developed whiskers.

These are no ordinary hairs. At least twice the width of body hair, whiskers grow from roots that are anchored in the skin three times as deep as the rest of the hair on the body. Out in the dangerous real world, the whiskers will vibrate in response to the slightest air current, helping the dog sense the proximity, shape, and size of objects it encounters. They also help protect the eyes and the rest of the face; a single faint touch on a whisker will cause the eyes to blink.

Those eyes, too, have been growing and changing. As we have seen, dogs' eyes have a number of specialized adaptations. For instance, the pupils can dilate a great deal to admit more light. But dogs also have other ways of improving their ability to frolic in the dark. In the more than three weeks since we last looked at them, the eyes have begun to develop a special layer behind the retina called the tapetum lucidum (or simply the tapetum). This mirrorlike lining of glistening cells reflects incoming light back to the retina to improve the dog's vision in faint light. Its reflecting material is what makes your dog's eyes acquire a ghostly blue glare in flash photographs—the same phenomenon you glimpse when your car's headlights pick up the eyes of a deer by the side of the road. The tapetum greatly increases the dog's ability to see in low light, but it also contributes to the reduced visual acuity discussed earlier. This will continue growing after birth; after all, it isn't needed in the womb.

Many animals have specialized, relatively stiff hairs, called vibrissae, that form on the face or, in some species, on the forepaws. They serve as tactile sensors, as when a cat squeezes through a tight space guided by its whiskers more than its eyes. Vibrissae have no nerves but are anchored in sensitive follicles that convey vibrations to the brain.

Not every dog has equally good vision. For one thing, breeds possess wildly different ranges of peripheral vision, depending upon how the eyes are situated in the head—flat in the front, like those of a pug, or close to the side of the head, like those of a greyhound.

We've discussed the mouth, nose, eyes, and even whiskers, but where are the teeth? The fetus has begun to grow its milk teeth, but they are still submerged in the gums. They won't erupt through the skin of the gums until at least three weeks after birth. Newborn animals that suckle for their nutrition don't need teeth; they could also harm the mother's teats. In contrast, animals that need to enter the world ready to gather and chew their own food are born with their teeth already in place.

Hair follicles are now producing sensitive whiskers.

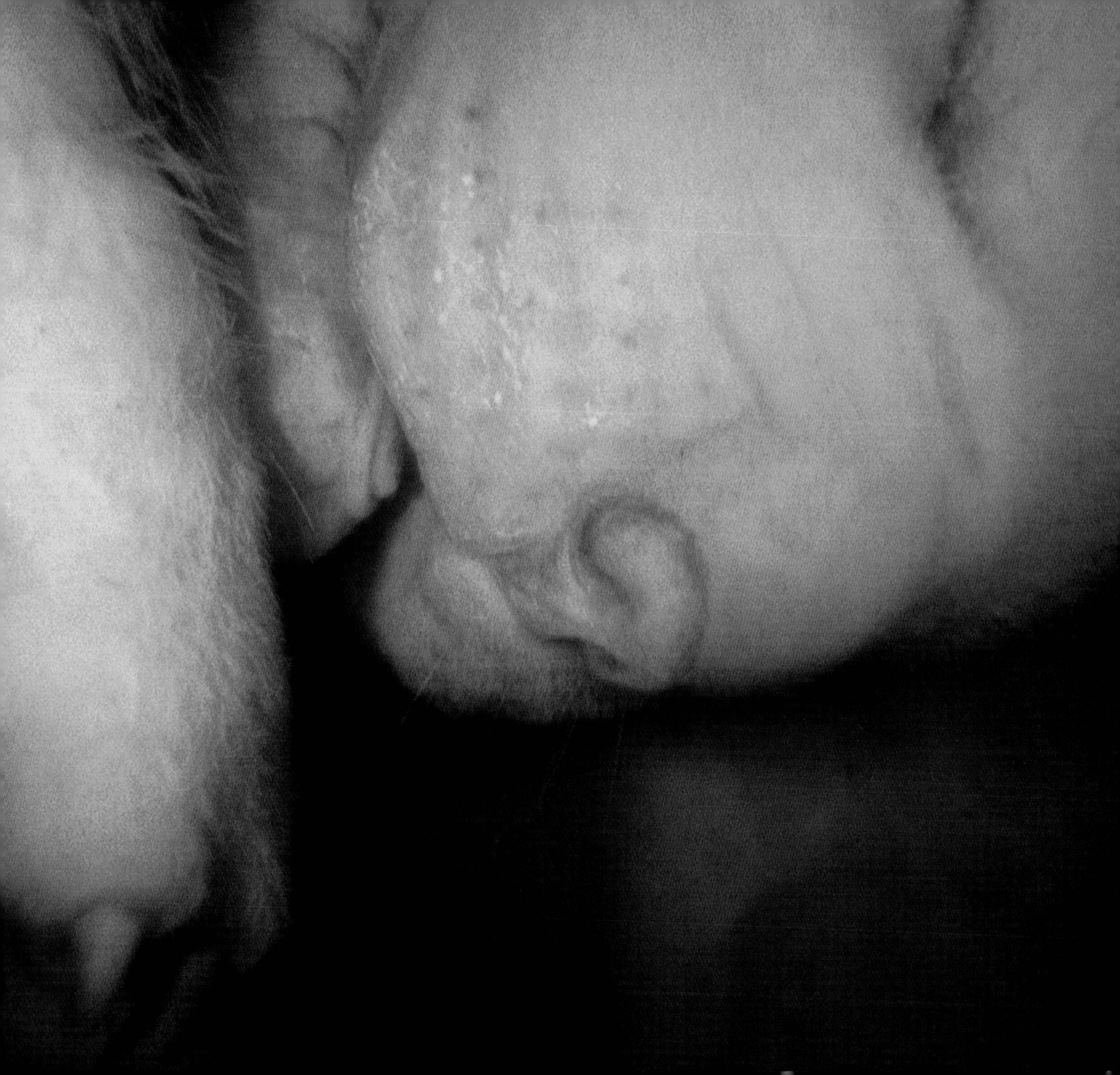

Ghosts of the Past

It is only two days until our golden retriever will whelp, as giving birth is called in reference to carnivores such as dogs, wolves, and bears. In the last few days of the pups' nine weeks in the womb, they grow rapidly. The mother's uterus is so crowded now that both she and the pups feel stressed and uncomfortable. The pressure against the uterine wall stretches blood vessels, causing them to transmit less blood through the umbilical cord—thus reducing the pups' oxygen supply.

This sounds like a risky situation, but declining oxygen levels help trigger important changes in both the mother and her offspring. Throughout their existence so far, the fetuses have been living like astronauts plugged into the reliable support system of the mother ship, growing peacefully while protected from a harsh, cold world. Now their heart and lungs must prepare to survive on their own. Before the pups can enter the world, however, their mother must prepare as well. The lowered oxygen level triggers her body to produce a round of different hormones. These initiate the birthing process by stimulating subtle, irregular contractions in the muscles of the uterus. Such movements prepare the birth canal by relaxing, lubricating, and dilating the uterus, vagina, and vulva.

Meanwhile, other hormones are triggering behavioral changes visible from the outside. An attentive pet owner will long since have provided extra bedding for the mother, preferably in an enclosed and covered place. Newspaper works well as bedding, because the mother can easily shred it as her hormones tell her to prepare a soft nest for the pups. If she has towels or rags in her bed, she will paw and bunch and attempt to shred these as well. Scientists theorize that this behavior reflects the whelping behavior of wolves in the wild. Dogs' ancestors evolved for millions of years before the few thousand in which they have lived with humans. For their homes, wolves dig burrows in hillsides or occupy caves or crevices. In his book *Dogwatching*, Desmond Morris argues that, because of this primordial instinct, a dog considers a house "a huge den with various entrances leading through tunnels into enlarged cavities. . . . All that is missing is the gently curving floor of the cavity itself." This last piece of the ancient nest the dog herself tries to create by pawing the "earth" beneath her and shredding her bed.

About a day before delivering their litter, a female typically stops eating. If she doesn't already have a secluded, burrowlike bed, she will seek out a private, enclosed space under a bed or in the corner of a closet. If she's left outdoors for long, she may try to dig a burrow.

Wild dogs are just as adaptable as their tennis-ball-fetching cousins. We find them all over the globe: dholes in Manipur, spotted hyenas in Namibia, red foxes in Tennessee. Maned wolves chase capybaras in Paraguay, and Australian dingoes still leave their tracks in billabong mud. American coyotes whose ancestors trailed Lakota hunting parties now steal garbage in suburban Atlanta.

Red wolf pups in the kind of den that a dog's nesting recalls

Get in Line Day 63

During the past 63 days—nine weeks—the embryos have grown into fetuses and the fetuses into pups. Through the amazing process that all animals undergo before birth, they have developed from a single cell into fully formed animals that comprise literally trillions of cooperating cells. The pups are ready to leave their mother's womb and make their own way in the world. Soon we will see how dolphins perform their extraordinary variations on this process underwater and how elephants grow in the womb despite the complications of their extreme size. Perhaps we find this mammalian drama so moving because we know that each of us undertook the same journey before we ever opened our eyes and saw our mother's face.

As birth draws near, breeders or veterinarians closely monitor the mother's temperature. In one of the first signs that birth is imminent, within 24 hours or so before the onset of labor, her temperature drops from an average of 100.4°F to 98.6°F. In dogs as in humans, birth varies enormously between individuals. The process is easier in both species if the mother is healthy and fit and housed in good conditions. Under ideal circumstances, a female's delivery of an entire litter may take only an hour. Different breeds whelp more slowly, and more difficult conditions may prolong labor even as long as a day and a half.

Now the mother starts to pant to cool her overheating body. As uterine contractions grow stronger and more regular, she lies down to ease the process of giving birth. Inside the womb the fetuses change position, because their longtime nestled posture doesn't line them up correctly to be born headfirst—although birth tailfirst is not uncommon. The contractions push the pups through the uterus toward the vagina, with the most powerful contractions needed to squeeze the pup's shoulders and head through the birth canal. This process dislodges each placenta from its connection to the uterine wall, enabling it to move and thus continue to protect the fetus that it has guarded for so long.

It is time for the mother to begin delivering her litter. The relative ease with which most mammals give birth reminds us that the pain experienced by many women results from the oversize skull of human newborns, which has enlarged to accommodate the complex brain of *Homo sapiens*. Most wild animals deliver their young without such complications. Dogs, however, suffer the burden of our endless tinkering with their breeding, as we have tweaked them like bonsai to bring forth the kind of new and improved dog we envision. The head of bulldog pups, for example, can be too large to pass through the birth canal. There are other breeding-related problems as well. Dalmatian mothers may give birth to more pups than they can feed, and Yorkshire terrier litters may be so large the pups have to be delivered by cesarean section. But most breeds deliver without much difficulty.

Equipped with fur and paws, the pups are ready for the world.

Unwrapped Birth

One at a time, the pups leave their home of two months. Each emerges still wrapped in a membrane or sac, although it may be torn, and soon each pup's placenta follows it into the world. The mother tears open the birth sac and licks the pup clean, stimulating it to begin breathing on its own and moving around. Then she nibbles through the umbilical cord about an inch from the pup's abdomen. At this spot, when the cord falls off, the pup will have a navel–a reminder of the physical link between generations, as well as a badge of membership in the vast clan of placental mammals, alongside dolphins and elephants and human beings.

The mother turns next to the placenta, which has one more task to perform. She eats it. Its still-active hormones stimulate the flow of milk from her teats and encourage delivery of the next pup in line. They arrive a few minutes apart, each slightly different from the others, as befits creatures grown from the genetic contributions of two individual parents. After the whole process is over, the placental hormones will help the uterus shrink back to normal size. Wild dogs have yet another reason to quickly eat the evidence of a fresh birth; they don't want predators to find a clue that vulnerable new lives are in the neighborhood.

Finally all nine pups are out in the world, unwrapped and nursing for the first time at their mother's teats, where their smart nostrils immediately led them. The tired mother watches over

The mother's licking stimulates the pups to breathe.

NOSE RE-SHAPING

her new family. Blind, damp, and defenseless, her pups clamber over each other in slow motion, already listening and sniffing and touching—beginning to employ the sophisticated sense organs that developed during gestation. Each pup's twitching nose draws air into brand-new lungs; its heart pumps blood through tiny limbs. Their ruffled fur is already beginning to dry.

The degree of development at birth varies enormously among mammals. Precocial animals, as the word's similarity to "precocious" indicates, are born (or hatched) relatively mature—at least mobile and alert, if not necessarily able to fend for themselves. Creatures born helpless are called altricial, meaning "requiring nourishment." Hares and rabbits are a good example. Hares remain in the womb an average of 11 days longer than rabbits, granting them a developmental head start. The jackrabbit of the western United States is precocial, its young born wide-eyed and furry and ready to leap away from danger. In contrast, the familiar cottontail rabbit of eastern North America is altricial, giving birth to hairless and vulnerable young, usually while safely hidden away in burrows underground.

Although furred, these golden retriever pups are otherwise altricial. It will be a week or ten days before their eyes open, and even then they will remain blind for a few more days. They must grow for two weeks before their ear canals open and for a few days longer before they begin barking. They will have been in the world a month before they start gamboling about and getting into trouble. Soon afterward, however, their hearing will mature to become far more acute than our own. Although the timing of the first mating varies depending upon breed and circumstances, both males and females will reach sexual maturity in less than a year, ready to begin the reproductive cycle all over again.

These dogs have a lot of growing left to do, but already they have come a long way. What a journey, from single cells contributed by their mother and father through the whole beautiful and mysterious process of growth in the womb. During the two months in which we have been observing them, their bodies have sculpted themselves from a barely animalian embryo to an obviously canine fetus to a furry miniature golden retriever. The pups are now independent animals with their own busy future ahead.

One journey is over and a new one has begun.

Emperor Penguin

Red Kangaroo

Part 2 Extraordinary

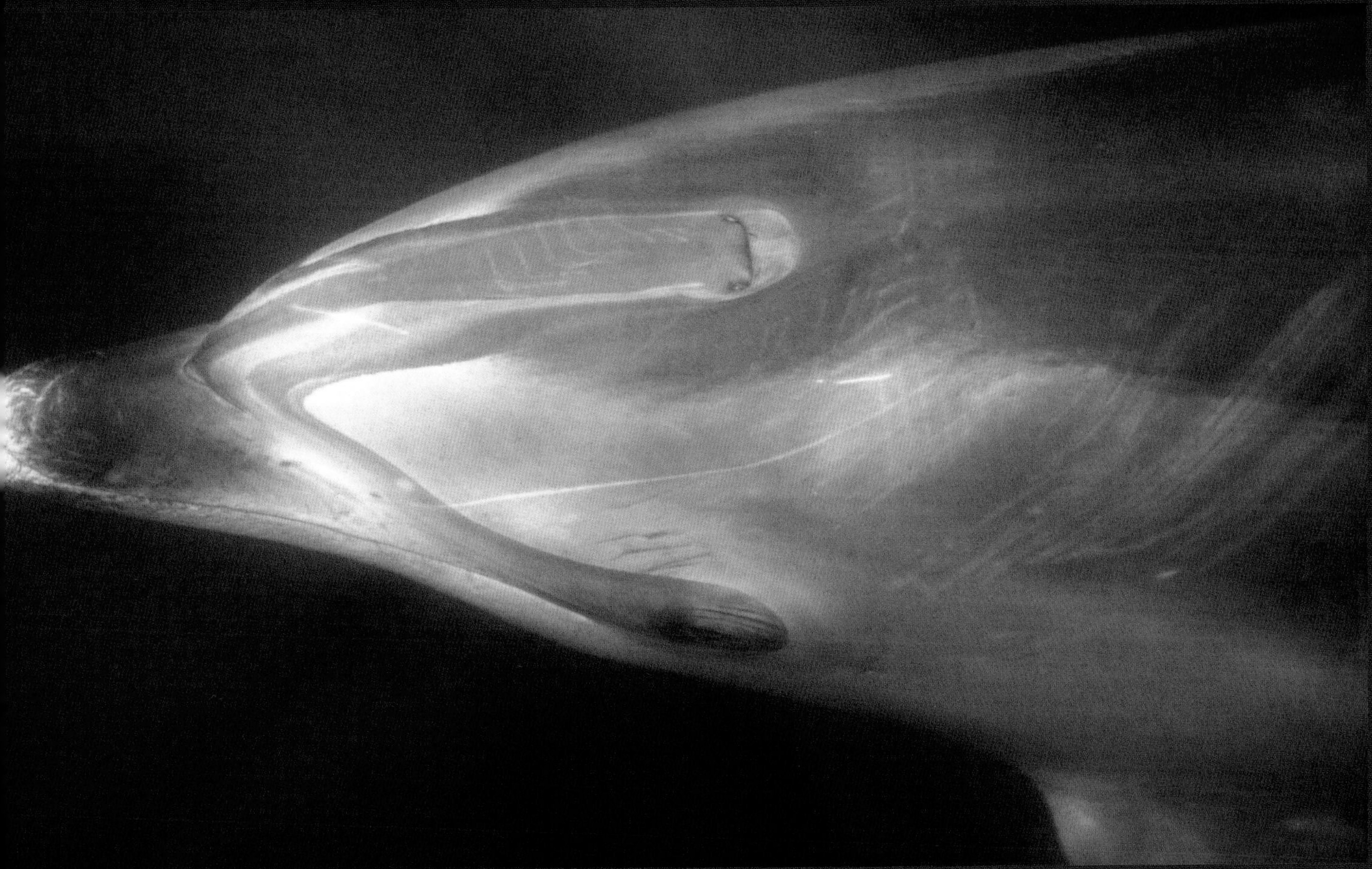

Bottlenose Dolphin

A Glee of Dolphins

In treating the dog as a familiar norm and the dolphin as extraordinary, we are demonstrating our human biases. The dog is familiar to us, but it's hardly a representative animal, having been reimagined and modified so many times in its relationship with humans. But no one could argue that dolphins are anything but extraordinary. It's true that, for some humans along coastlines and islands around the world, dolphins are not uncommon—but they are never ordinary.

Different animals appeal to us for different reasons, depending upon everything from their behavior to our cultural background. Some zoogoers can't resist tigers, while other people head straight for the pandas; many of us divide the world into dog people and cat people. But surely only a hopeless curmudgeon could resist a dolphin. Dolphins inspire joy. From their vivacious antics to their resting expression shaped like a human smile, we find them irresistible. When the British poet C. Day Lewis wanted a collective noun for these animals, he naturally thought of a "glee" of dolphins.

We have always claimed playfulness as their identifying characteristic, just as we have long thought of them as friends. "Dolphins are not afraid of humans as something alien," wrote the Roman encyclopedist Pliny the Elder almost 2,000 years ago, "but come to meet vessels at sea and play and leap around them; they try to race ships and overtake them even when they are under full sail." Millennia of encounters have only burnished dolphins' reputation among human beings, especially now that scientists have studied them closely.

In Part 2, we will follow a bottlenose dolphin from mating to conception to birth. Of 30-odd species of dolphins, the bottlenose, *Tursiops truncatus,* is best known; it's the one that most of us visualize when we hear the word "dolphin." Flipper, the aquatic equivalent of Lassie in the 1960s TV series, was a bottlenose, as were the stars of the George C. Scott thriller *Day of the Dolphin.* As both its common and scientific names indicate, the bottlenose's snout—called a beak—is unusually short for a dolphin.

A handful of dolphins live in fresh water. The largest, the bouto—famous as the pink dolphin—lives in the Orinoco and Amazon river systems in South America. Another freshwater species reminds us that the physical appearance of the bottlenose may be more important in our affection for it than we realize. Its cousin, the baiji, the Chinese river dolphin, has a long, narrow snout and beady little eyes. Biologists think that its less appealing face has contributed to the lack of concern over its drastic decline in recent decades, while the bottlenose has become the face of ocean conservation.

Dolphins escorting a cargo ship in the Strait of Gibraltar

***previous pages*: Dolphin society is lively and complex.**

Social Sexuality

We regard dolphins as extraordinary, but they resemble an animal so common we don't even notice its behavior as animalian—*Homo sapiens*. Like us, dolphins are gregarious, inquisitive, and affectionate, and engage in sexual activity more often than necessary to maintain the population. Dolphins don't just mate to reproduce. They mate because they enjoy it.

"Enjoy" may seem an unscientific term, so perhaps we ought to say that, according to biologists, the physical pleasure derived from sexual stimulation improves social bonds between the participants. Females and males reach sexual maturity at about the same age—between ten and twelve for females, with males lagging a bit. Although females ovulate no more than seven times a year (usually less often and sometimes as seldom as twice annually), they are ready to mate 365 days a year.

Perhaps even more surprising, dolphins' sexual activity isn't limited to paired males and females. Adult males may attempt to mate with pregnant females, with females still guarding a young calf, with calves that are too young to reproduce, and even with other males. Females may engage in sexual antics with their own or other calves. Juvenile play apparently includes mutual genital stimulation. Dolphins have been observed stimulating each other's genital region with their echolocation sonar.

Why did such varied sexuality evolve? One scientific guess: Dolphins may do this in part to demonstrate to their young the importance of social cooperation. Social bonds are crucial. A dolphin pod is a loosely defined entity, flexible and temporary, comprising an average of five members that may or may not be related. Like us, they cooperate socially largely because doing so benefits them individually and creates a safer community for their offspring. Scientists also propose other rationales for this promiscuous system, and as usual the answer most likely involves several factors. For example, random mating results in uncertain paternity. In many species, males kill other males' offspring; uncertain paternity may circumvent this risk. Sexual bonding may also facilitate allomaternal behavior, in which one female cares for another's young when necessary. A number of other creatures, from squirrel monkeys to wolves, have also been observed "babysitting" (see page 34).

But it is also important to remember that nonreproductive sexuality is not uncommon among animals. Humans have always had a list of traits we found interesting in ourselves and therefore imagined to be unique—even to the point that we claimed them as what scientists call species-specific traits. Tool use, humor, warfare, politics, more sex than necessary—we have proudly claimed all of these activities for ourselves, only to find later that we had again been deluded by our yearning for uniqueness.

Gregarious and affectionate. No wonder we like them.

Hurry! Week 1: Part 1

Although dolphins have few natural predators, those few are fierce—killer whales and large sharks such as the great white, tiger, and bull. In studies off the coast of both Australia and Florida, roughly a third of all dolphins that scientists observed bore bite scars from attacks by sharks and killer whales. This number was true of both sexes and all age groups. Scientists don't know yet whether these scars record a predator's attempt to kill a dolphin for food or attacks that are meant to warn dolphins away because they are regarded as rivals for the same fish. Despite our mental image of them as lovable, dolphins are highly efficient predators themselves, and would certainly gobble up their share of the local buffet.

Just as the threat of predation may have nudged the evolution of dolphin pods—the safety-in-numbers rule—so may it have led to their mating routine. Scientists estimate that only 10 percent of dolphins' sexual activity involves true intercourse; most of it is foreplay. Usually mating begins with the kind of playful genital and nongenital touching that is common even among the pod members that aren't actually warming up for mating. The flirting pair chase each other and rake each other's skin with their teeth. One may float lazily on its side and permit itself to sink downward in the water while the other cavorts around it.

Dolphins don't exhibit what scientists call secondary sexual characteristics, which means that the sexes don't look different, unlike deer and lions and humans. Females have a single slit on the lower abdomen that houses both genitals and anus; males have two slits, one for each. We can't tell which we're looking at without seeing an actual erection or intercourse or a close-up of these hidden slits. Most male animals carry their testicles and penis outside the body, because mammalian body temperature is too warm for sperm. Yet a marine predator such as the dolphin had to evolve a streamlined shape. Their genitals gradually moved inside the body, to be cooled by a specialized network of blood vessels that we will examine as they develop.

Apparently males can achieve erection at will, without physical stimulation. And unlike dogs, which may spend an hour in the tie stage (see page 26), dolphins mate in a few seconds and quickly separate. With the risk of attack so high, it makes sense to get the actual mating over to limit the period of lowered defenses. Distraction can be fatal. As if mating underwater were not complicated enough, mating in seawater is even more of a test, because seawater kills sperm. Therefore, as the male inserts his penis into the female's vagina, her muscles lock around it and create a water-tight seal to keep salty water away from the sperm.

Fear of predators makes dolphins mate quickly.

A Competitive Nature Week 1: Part 2

Every aspect of an animal's life reveals wonderfully strange adaptations that provide clues to its evolutionary heritage. During the few seconds of actual intercourse, for example, the male dolphin ejaculates as much as two and a half ounces of fluid, but most of it has a low percentage of actual semen. The amount may be high to flush out the semen of competing males.

Another hint of sperm competition is that, when the semen does arrive, it's highly potent. Male dolphins have the disproportionately large testes and large quantities of sperm often seen in species that engage in what biologists call promiscuous mating behavior. Competitive sperm production might give one particular male the edge over his possibly numerous rivals. Their semen can contain as many as 10 million sperm per ounce—the highest known concentration in the world, 15 times that of a healthy male human.

Even when mating is over, the female's body doesn't relax its vigilance against the challenges of living in salt water. As the male withdraws his penis, his movements trigger contractions in the female's pseudocervix. This marine adaptation, which is unique to cetaceans and manatees, consists of annular (ring-shaped) folds outside the actual cervix. The pseudocervix's bundled muscles clamp shut against salt water to protect the sperm. Apparently its muscular contractions also begin the process of moving the sperm toward her eggs. First they have to pass through another barrier between the pseudocervix and the actual cervix—a mucous plug that further protects the uterus. The female's body deposits the sperm here, where it is safe from salt water.

Dolphins are not the only creatures that have evolved such outrageous reproductive ploys. In the parched desert and scrubland of the Australian interior, red kangaroos face their own gauntlet of challenges from both environment and rivals. The largest marsupial on Earth, *Macropus rufus* can weigh 200 pounds and reach five feet in length. When they mate, the male kangaroo's semen begins to coalesce very shortly after entering the female's vagina. It forms a thick, rubbery bung that completely blocks the passage. Although it will permit this particular male's sperm to pass through, scientists think that it guards against that of rivals.

"We will now discuss in a little more detail the struggle for existence."

Charles Darwin

The red kangaroo has also evolved another, equally specialized, method of reproductive insurance. The female's vagina divides into two channels, each leading to a separate uterus and ovary. Half the sperm that get through the semen plug travel in one direction, and half go the other way. Only one team will find an egg, however, because only a single uterus is fertile at any one time.

The pseudocervix protects sperm from seawater.

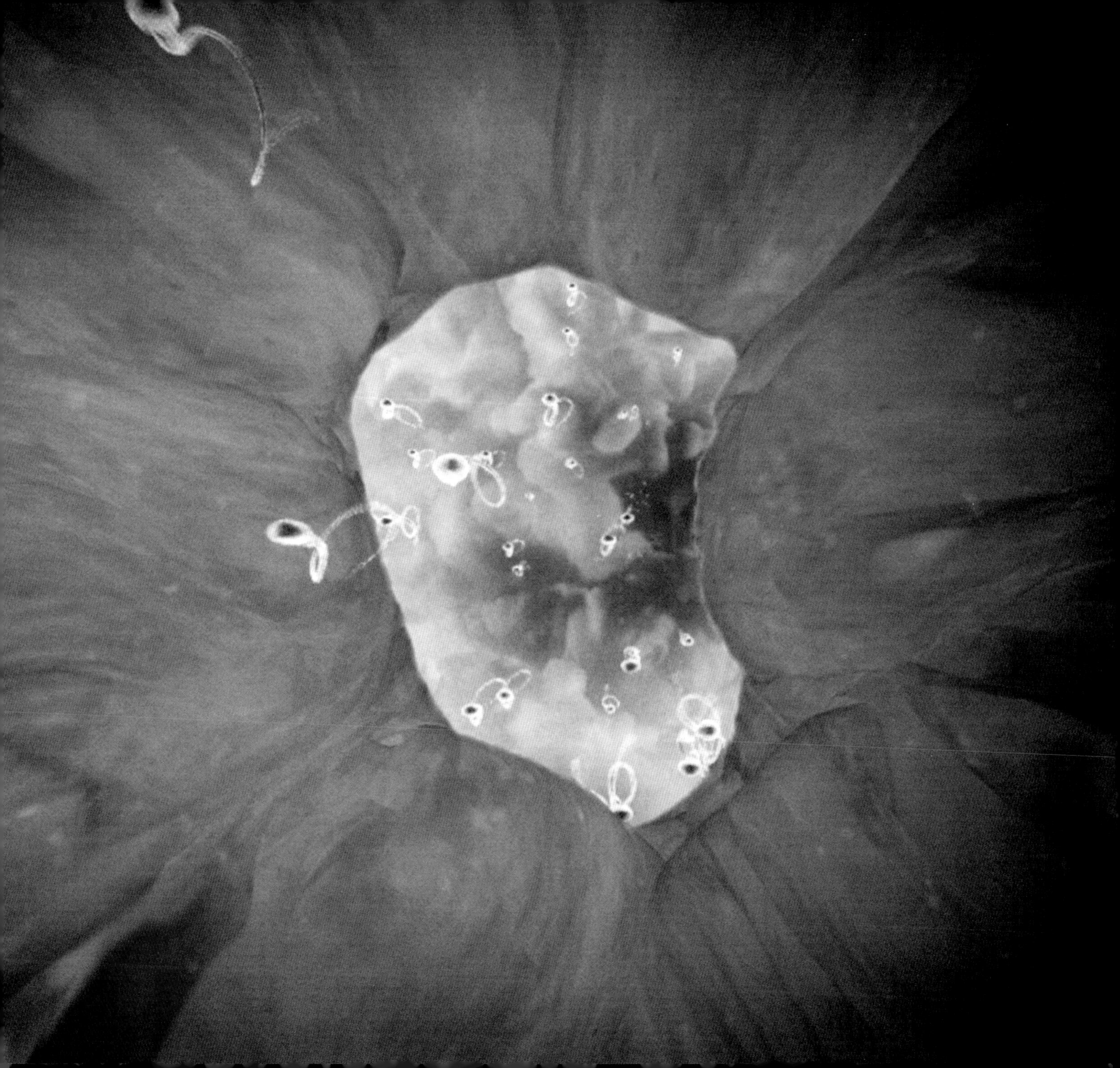

The Family Tree

Remarkable changes are taking place in the embryo, changes that may provide a glimpse into its distant past. Three to four weeks after conception, when the embryo is only half an inch long, limb buds begin to appear. Near the embryo's roughly modeled head, the early shaping of the flippers has begun, but these limb buds are different, because they aren't forming flippers; they're beginning to form hind limbs. Over the next two weeks, these buds will grow larger and then retract and finally disappear completely, as if they had never existed.

Modern dolphins, of course, show no sign of hind limbs on their streamlined body. In fact, the absence of hind limbs is an identifying characteristic of the mammalian order Cetacea (whales, porpoises, and dolphins), distinguishing it, for example, from semi-aquatic sea lions and walruses, whose hind flippers are modified legs. In the photograph on page 12, you can see what can happen to a dolphin when the usual chemical signals in the womb go astray and hind limb buds do not receive their instructions to reverse development and disappear.

So why does the embryo start sprouting legs and then seem to change its mind? As we have seen in dolphins, as in other creatures, including human beings, the development of the individual in the womb reflects aspects of its group's evolutionary history. These leglike limb buds hint that dolphins evolved from four-legged terrestrial ancestors. Gradually, researchers have narrowed the field of possibilities to a wolf-size carnivorous mammal called *Pakicetus*. It lived roughly 52 million years ago, early in the Eocene epoch.

The study of evolutionary kinship between groups of organisms, called phylogenetics, indicates that *Pakicetus* is closely related to cows and hippos, while other studies point to bearlike ancestors for seals and walruses. Fossils from the Eocene epoch document many stages of mammals' progression from terrestrial to aquatic. Before the discovery of *Pakicetus,* however, the most primitive mammal known was obviously amphibious and lived in coastal regions. *Pakicetus* takes the ancestry back onto land. It may have hunted in rivers or along seashores or both, but the evidence indicates that it definitely hunted in water.

The link between *Pakicetus* and cetaceans—and the reason for thinking that it pursued aquatic prey—is hidden in the structure of its ears. *Pakicetus* possessed a trait that scientists have never found in any creatures other than cetaceans: an adaptation in the inner ear for hearing underwater. Over millions of years, as habitats change, animals change with them, migrating from the ground to the air, as in the case of bats, or in the opposite direction, as penguins did; and from land to water and back again. As we will see in Part 3, the ancestors of elephants once may have lived in the ocean.

A reconstruction of *Pakicetus*, possibly dolphins' ancestor

Inside or Out?

As dogs, dolphins, and elephants demonstrate, mammals vary as much as any other group of animals on Earth in their gestation and their level of development at birth. Four weeks into the pregnancy, the dolphin fetus is still hidden away inside the uterus, growing and changing. It is only just now visible to ultrasound, and few actual dolphinlike details are visible, other than the tiny nostrils that have appeared. These will now slowly begin moving toward what will be the top of the head, to form the blowhole.

By this time, the golden retriever fetuses were already halfway to birth, but this dolphin has only begun its journey. After the same amount of time, many other animals are also much further along in development. The variety of gestation periods among animals reminds us that each species has achieved its own balance through a long history of adaptation to the needs of lifestyle and environment. In birds, for example, you won't find anything like the cozy human or canine womb. Such a luxury would be excess baggage in ultra-lightweight creatures whose hollow bones and airy feathers have evolved to provide strength and durability without adding much cargo to the flight load. Emperor penguins (*Aptenodytes forsteri*) lost the ability to fly, but they kept most other bird traits, including the need to avoid excess weight. Therefore, as quickly as possible, they must expel their embryonic young into the cold, unfriendly world as those familiar but extraordinary balls of protein and potential—eggs.

During only 24 hours of frantic biochemistry after mating, a female Emperor penguin's body creates a huge egg. The walls of the oviduct spin a nutrient-rich yolk and coat its surface with albumen, a protein-based fluid better known as egg white. Then the egg's surface begins to harden as the body applies to it a hard coat of keratin, the same versatile protein that forms a dog's fur and an elephant's toenails and your own eyebrows. A last but vital step is a layer of crystals of calcium that coalesce on the surface, drawn from within the egg's own calcium-rich fluid to form the smooth, porous, and impressively tough shell.

Having depleted her own nutrient resources in crafting the egg and carrying it for a month, the female penguin is exhausted. Therefore, she turns the egg over to the male while she goes off to fish and renew her energy. Thanks to popular movies and documentaries, even people a long way from Antarctica are familiar with what happens next. The male shuffles around, carrying the egg on his fat comical feet, in a warm pouch under his abdomen, so that it is protected from both cold wind and the icy ground of the Antarctic. For two long months, the male hangs out with other eggsitting dads until his offspring hatches and the female returns.

Meanwhile, in the opposite end of Earth's spectrum of harsh environments—the hot Australian outback—the same period of time is just as busy in the life of the red kangaroo. Unlike dog and dolphin eggs, more like those of birds, a kangaroo egg is coated with keratin when it enters the uterus. Instead of imbedding in the uterine wall and forming an umbilical cord and placenta,

A male penguin carries the egg during the female's absence.

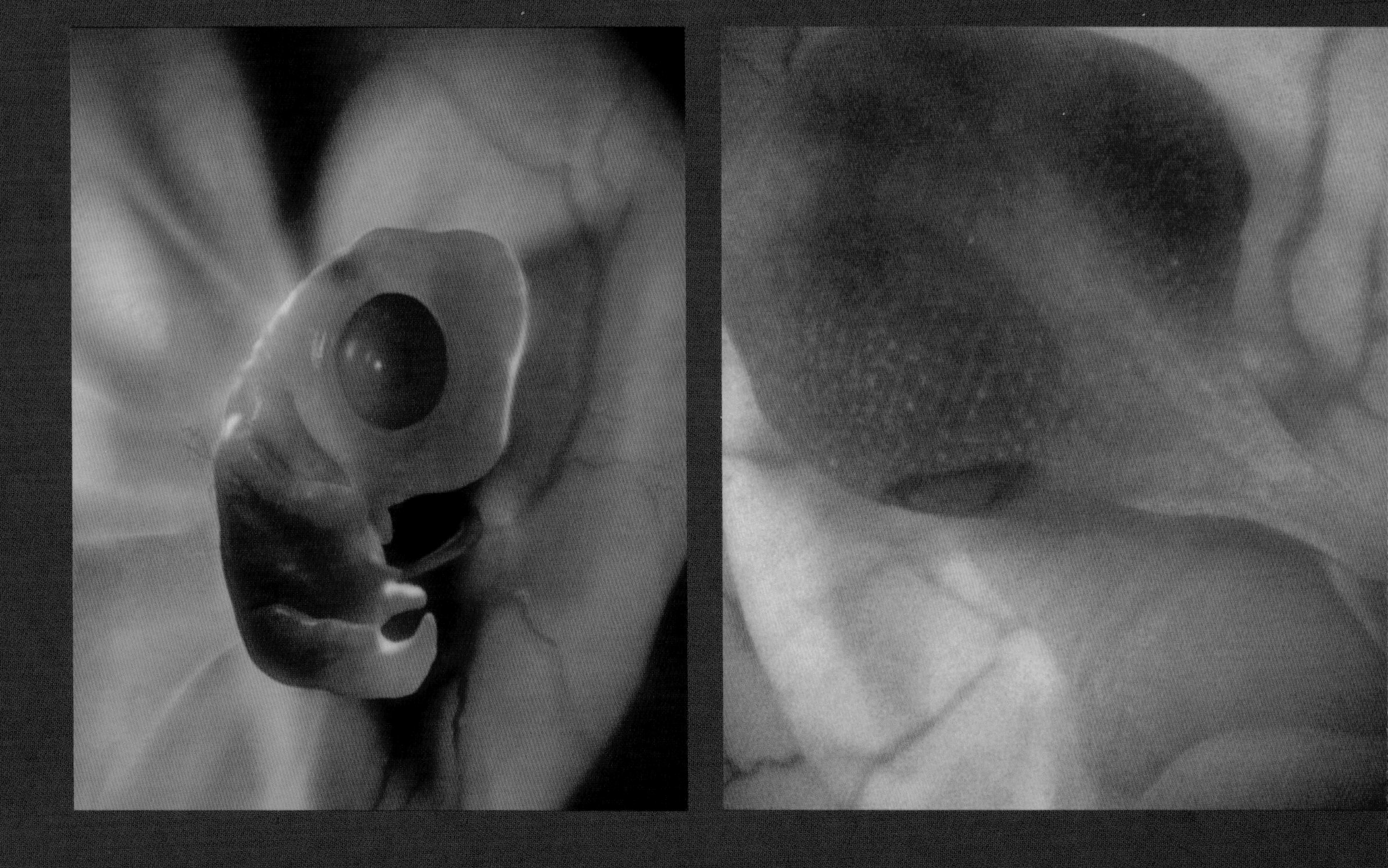

Inside the egg, this embryo is not quite a week old. The penguin's amazing egg is the key to its survival on the Antarctic ice. Carrying a single egg on their feet, penguins don't have to have a nest.

The egg, which is as large as an orange and can weigh up to a pound, is filled with nutrients that nourish the growing embryo. After three weeks, the embryo clearly shows beak and eyes, as well as the dimpled skin marking where feathers will soon grow.

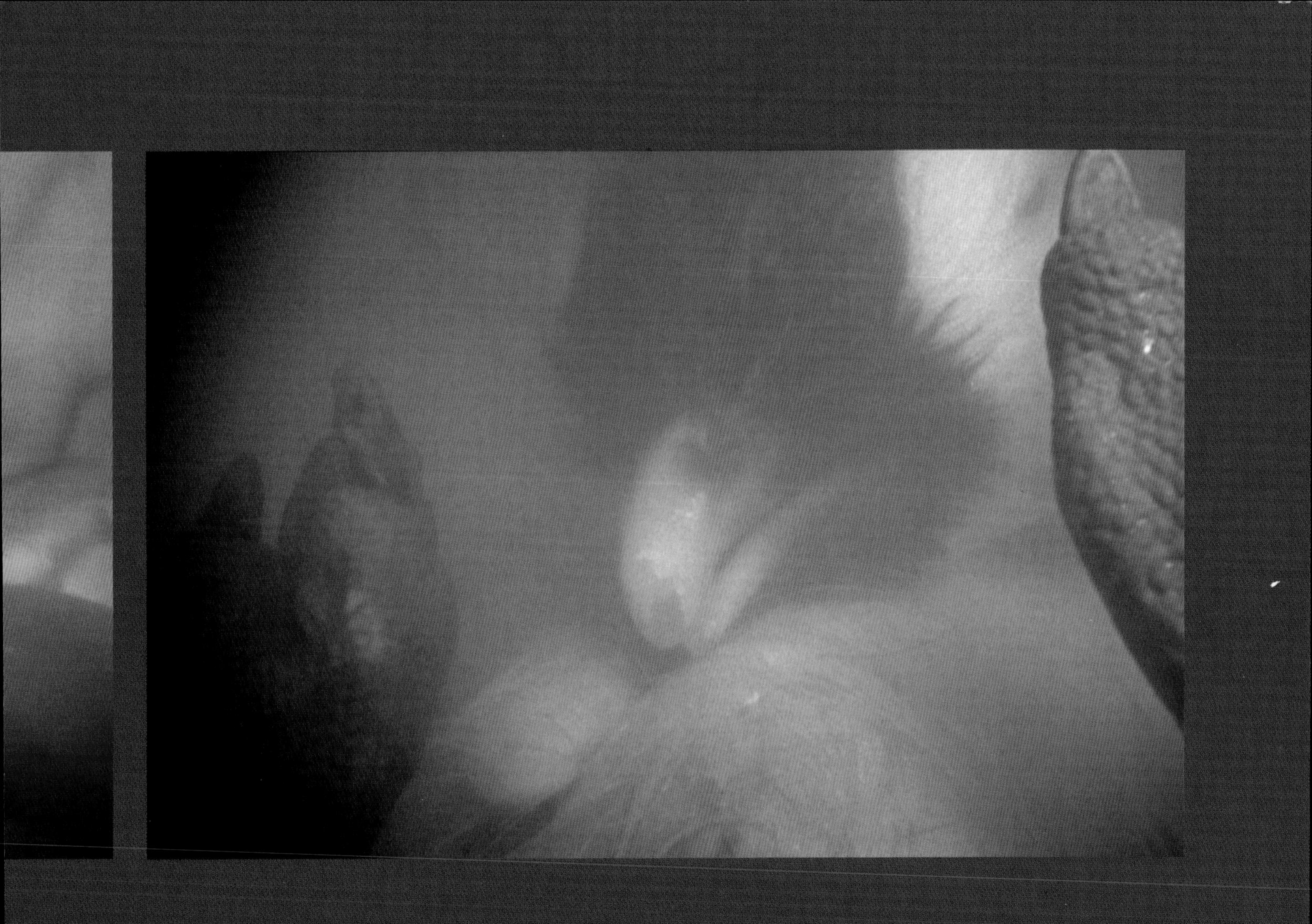

the egg is coated in a shell-like membrane that isolates it from the uterine fluid in which it floats. This lack of prenatal connection with the mother's system of nutrient supply is the key distinction between marsupials and other mammals.

"We find a Strange Animal among us . . . . An animal who carries her family about with her in her pocket!"

A. A. Milne, *Winnie-the-Pooh*

It also means that the egg can survive only a short time before needing to acquire a new means of sustenance. For a single month, it floats in the rich fluid, which resembles the albumen of the penguin's egg. Inside the kangaroo egg, the embryo's body is developing in curious ways. During this brief time, it predicts through its development which resources it will need soon after leaving the womb. The eyes won't be required immediately, so their development lags behind; likewise the hind legs, which barely resolve into legs at all before birth. But the inner ear's channels and the nasal passages form early, just as the forearms quickly grow disproportionately large and even develop a tiny claw on each finger. Clearly this embryo will soon need to orient itself, use its nose, and grasp something.

It does. Only 30 days after conception, the fetus releases the hormone cortisol into its liquid environment, triggering the uterus to give birth to a jellybean-size toy kangaroo called a joey. While her internal ecosystem is suffused with hormones, the mother licks a path along her own abdomen to lay a scent trail for her young to follow from birthplace to nursery. Like all newborns, the joey desperately needs nourishment. Immediately, its advanced nostrils pick up scent clues; its ears' balance organs orient it in relation to up and down; and the oversize arms grasp the mother's fur and begin life on Earth with an urgent journey. It crawls upward, struggling like a mountain climber, for the three suspenseful minutes it takes to crawl these few inches.

Then it finds its mother's pouch and latches onto a teat that expands to fill the baby's mouth and help it hold on. Like all marsupials, the red kangaroo has replaced development in the womb with an extended period of feeding on specialized and highly nutritious milk. The joey finds its mother's pouch and settles down for months of sleeping and eating and, eventually, peering out at the world. In another curious adaptation, the inch-long joey is bright red, because its blood vessels are close to the surface of the skin. Scientists think that this situation enables the joey's body to absorb oxygen directly from the air, to compensate for its still barely developed lungs.

At this point, when a penguin chick would be tapping its way out of its shell from the inside and emerging into the world rumpled and hungry, when a kangaroo joey would be hurrying through a jungle of body fur to reach its mother's teats, our very young dolphin embryo sleeps on in the carefully maintained environment of its mother's womb.

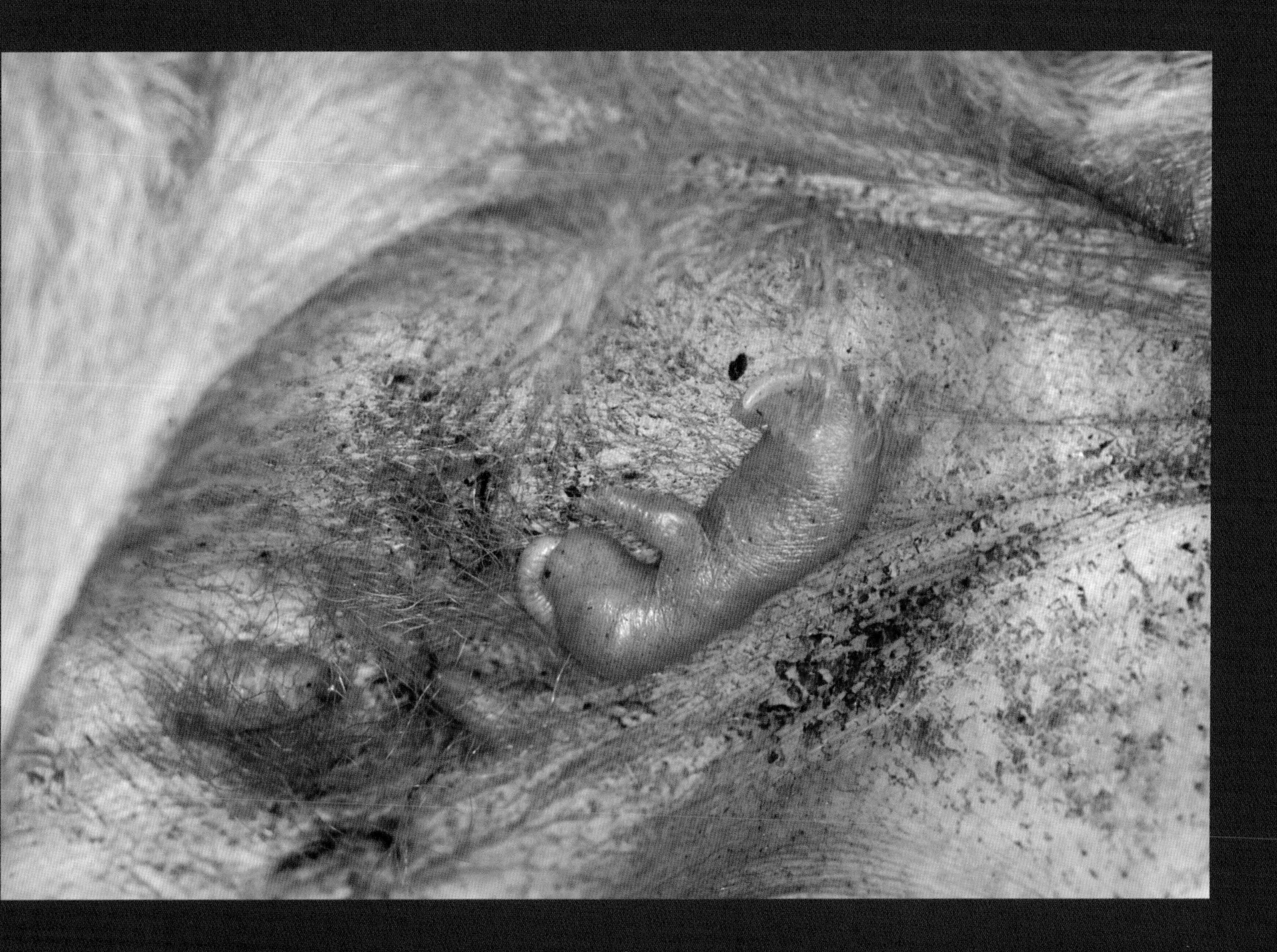

Kangaroos give birth early and complete the job in the pouch.

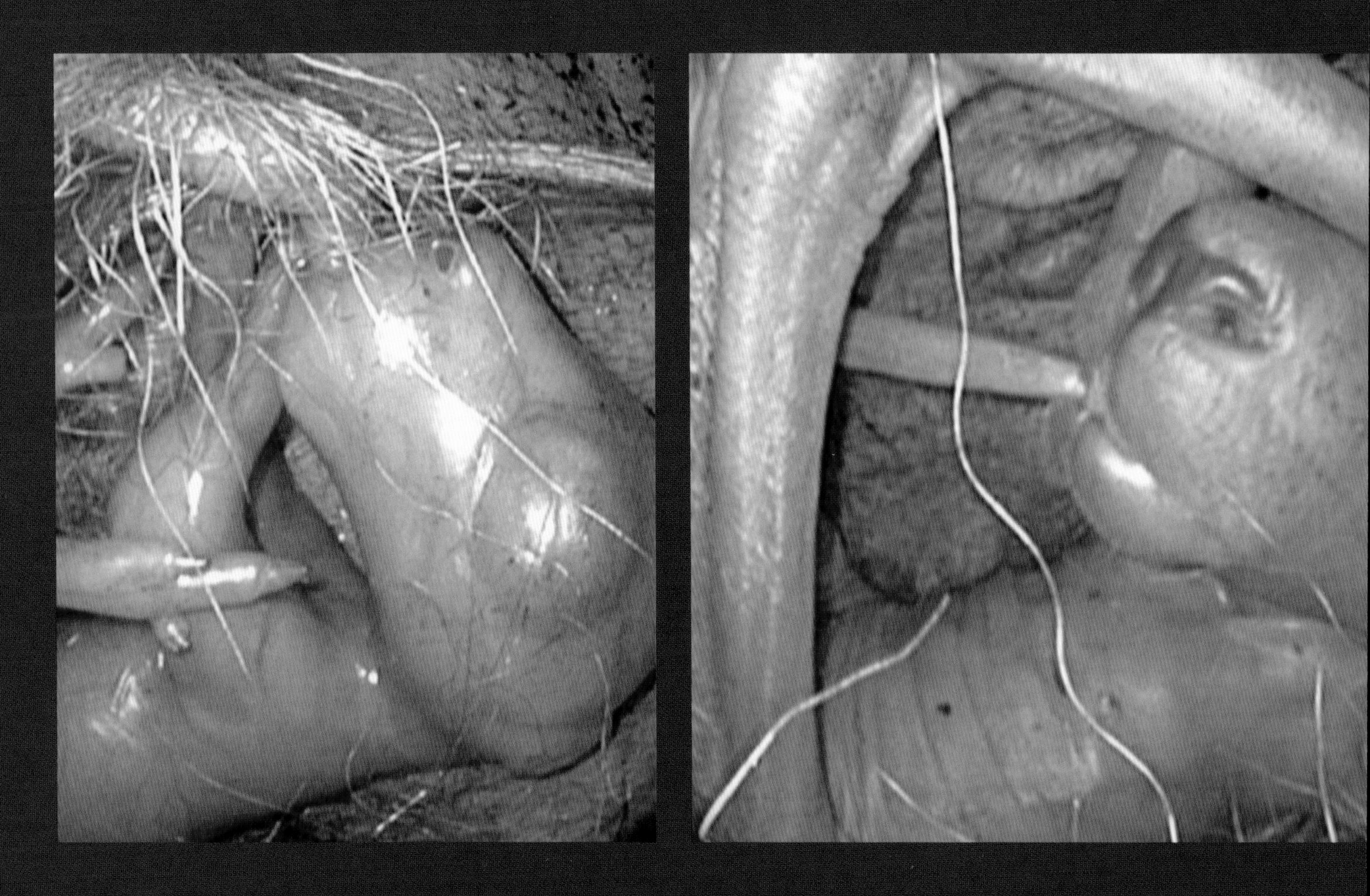

After ten weeks in the pouch, the joey still has a long way to go to even look like a kangaroo. Its mother's rich milk, which will change its mixture of proteins over time, completes the growth process begun in the womb.

Twenty weeks after it scurried up its mother's abdomen to hide out in her pocket, the joey is still firmly attached to its mother's teat, receiving all the necessary nutrients from her. It has yet to even open an eye.

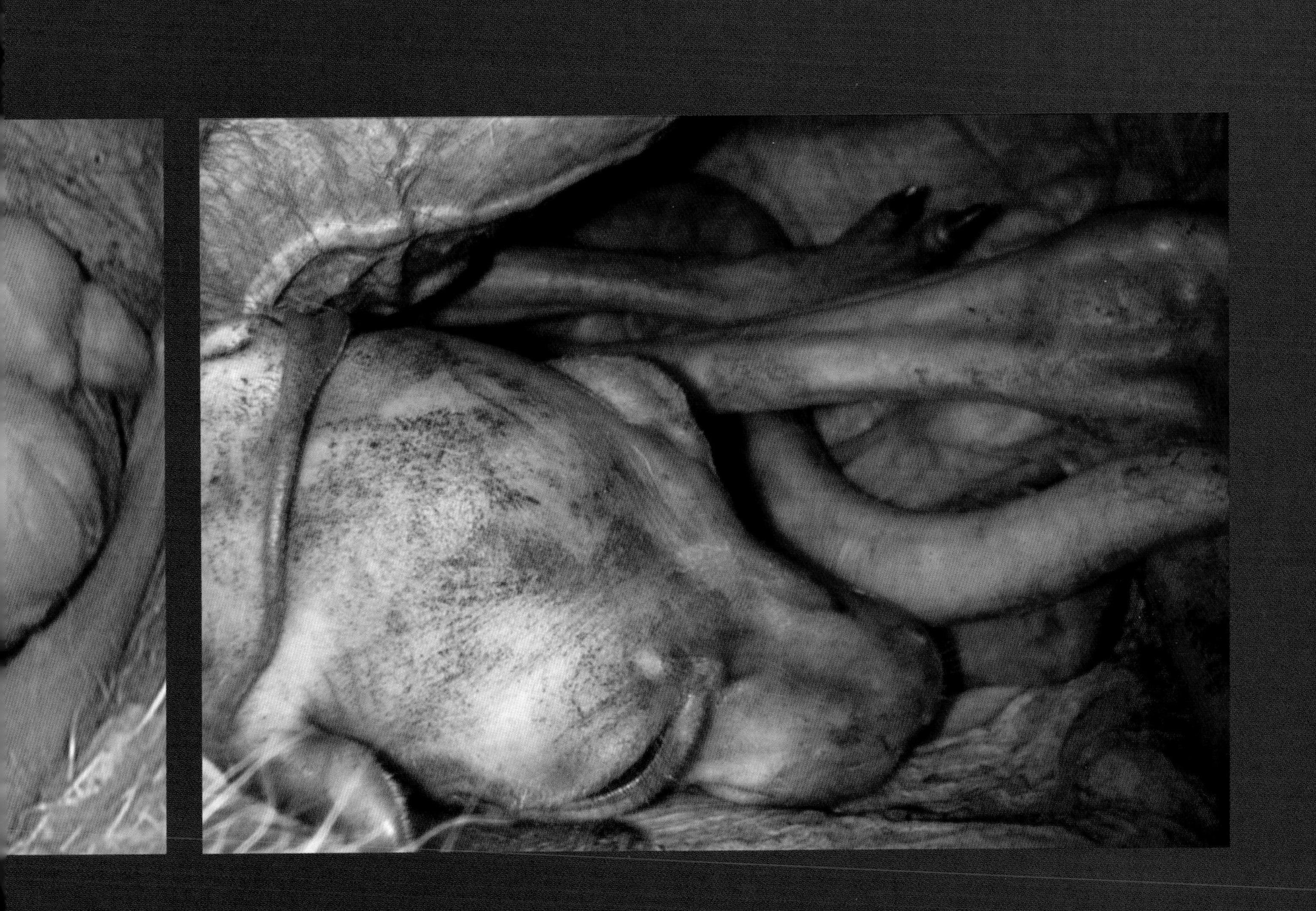

After six months in the pouch, the joey is the size of a squirrel. With its ungainly body folded up like a puppet's, it has three months to go before leaving the pouch for good. As a male kangaroo, the joey would be called a boomer; a female would be a flyer.

Hold Your Breath Week 9

The move from land to water that dolphins' ancestors undertook placed harsh demands on future generations. A body that originally evolved on land requires many adaptations to function underwater; imagine trying to turn your car into a submarine.

The most urgent question is how to breathe. Now, at nine weeks after conception, the nostrils that were barely visible a month ago have completed their climb to the top of the embryo's head and are merging to form the blowhole. Eventually, before birth, a flap of muscle will form a watertight seal over it. To breathe, the newborn dolphin will surface and open this flap and exhale forcefully to clear away water—blowing air up to 100 miles an hour. Then it will take a deep breath and close the flap to guard its air supply as it again sinks beneath the waves. The "blow" that we see is warmed water vapor meeting cooler air, along with seawater that was pushed ahead as the dolphin began to exhale before reaching the surface.

Dolphins have impressive lungs, but they can't keep up with their much larger cousins, the whales. Toothed whales such as the sperm whale can hold their breath for up to 80 minutes during their dives, which regularly go as deep as 2,000 feet, and often much deeper.

Dolphins never develop vocal cords because they have other ways of communicating. Just under the blowhole are air sacs that produce dolphins' distinctive clicks and whistles. And the blowhole is not even the most impressive piece of equipment in dolphins' respiratory system. We have the pleasure of glimpsing dolphins on the surface of the ocean because, although a dive may last up to eight or ten minutes if necessary, normally the adults breathe two or three times a minute. With every breath, they exchange more than 80 percent of the air in their lungs. Human beings, in contrast, must get by on a much less efficient 17 percent air exchange.

There is another major difference between our method of breathing and theirs. We are involuntary breathers, with no attention required for oxygen to reliably enter the body and get processed into the respiratory and circulatory system. Our breathing continues when we're asleep, as is true for dogs and most other mammals. Dolphins, in contrast, are voluntary breathers. To inhale, they must swim up to the surface, open the blowhole flap, and consciously exhale and inhale. At first glance, this method doesn't seem too difficult. But it doesn't just mean that dolphins must give the same kind of absent-minded attention to breathing that we give to speaking or walking. Conscious breathing requires that their brain must always be awake enough to bring in oxygen. Studies reveal that, at most, sleep in dolphins means that one brain hemisphere is active while the other rests. In perhaps the strangest of all their adaptations to life in the sea, dolphins never truly sleep.

The nostrils have merged to form the blowhole.

following pages: **The dolphin's signature shape is now developing.**

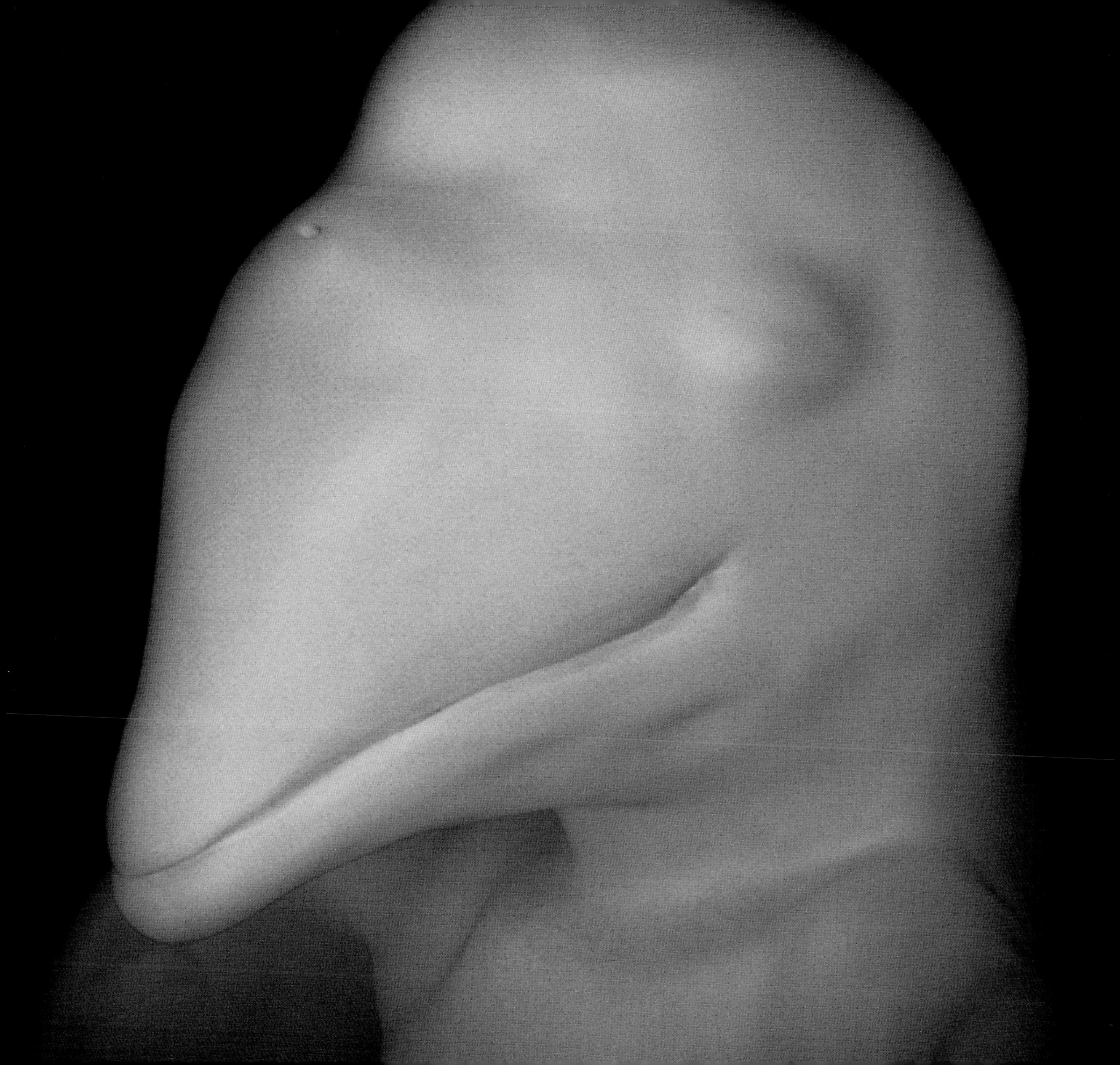

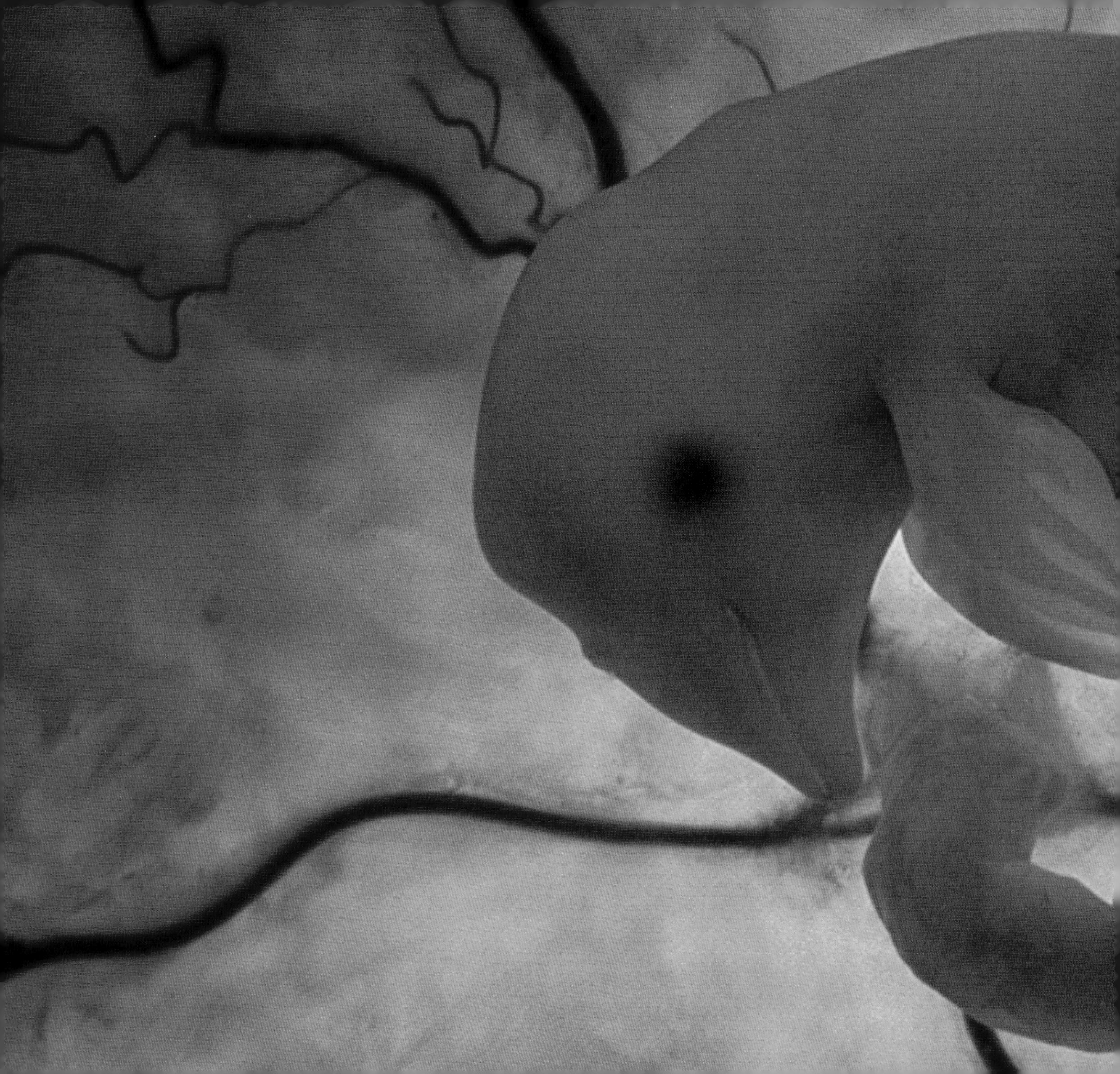

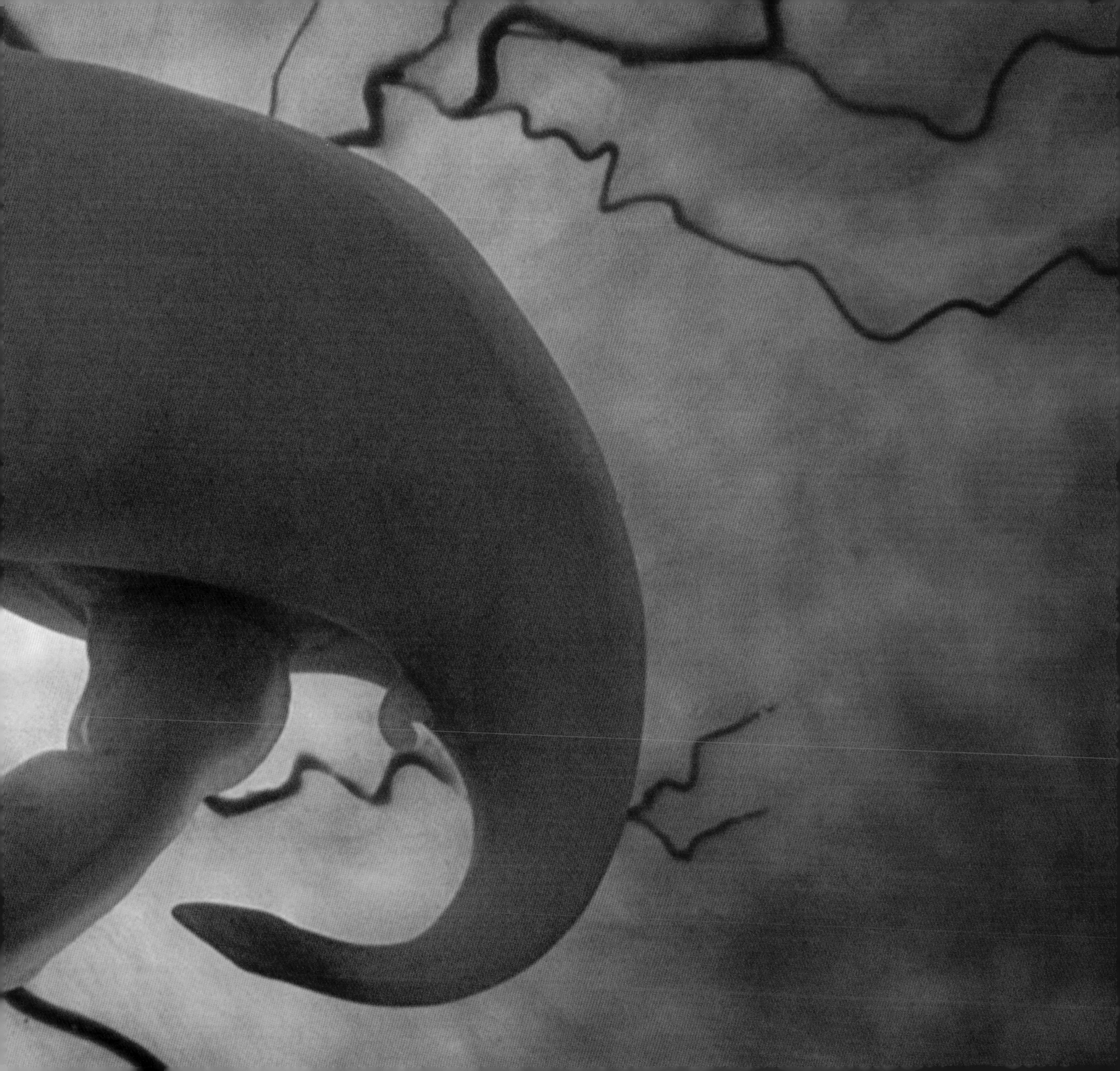

Walking on Water

By nine weeks after conception, the fetus is starting to look like a dolphin rather than like some generic animal shape. Specialists call the dolphin's configuration fusiform, meaning shaped like a spindle, wider in the middle and narrow at both ends to reduce drag. This shape will help the adult dolphin to swim close to 22 miles an hour, four times faster than an Olympic gold medalist. Increasingly streamlined and elegant, the fetus is already practicing what will be its primary activity after birth—swimming. It moves around in the amniotic fluid.

"If you swim in the sea, sooner or later you're going to end up looking like a fish."

Simon Ings

The flippers are forming, and so is the tail. Each lobe of the tail is called a fluke, but many people use the word as a synonym for tail. It will need to do more than just push the dolphin through water. Its muscles will be powerful enough to enable the dolphin to stand upright on its tail fin and propel itself across the surface of the ocean, as if this perpetually smirking mammal could top the rest of its popular act by walking on water. The tail can also propel the dolphin into those 15-foot-high leaps that look irresistibly joyous to us but probably have the practical purpose indicated by scientists' nickname for the behavior, spy-hopping.

In shallow seas, dolphins use these tail muscles to catch prey. In summer many dolphins hunt in water less than four feet deep, where fish are easily located and chased. A violent thrust of the tail can send fish flying through the air, stunning or killing them. We think of the dolphin as gentle and playful, but such on-board weaponry testifies to its efficiency as a predator and to its defenses against its own few enemies.

By 11 weeks, the pectoral fins—the flippers—have formed, not far behind the head and well in front of the big dorsal fin. With these fins, the adult will be able to steer and, with the help of the tail flukes, come to a stop. When a dolphin treads water or leaps into the air, its momentary upright pose reveals that the fins look like flat, little arms. This is precisely what they are (see page 90). If you examine an x-ray of a dolphin's pectoral fins, or if you even glance at a dolphin skeleton in a museum, you may be surprised to find how much the bones inside resemble your own arm and hand: ball-and-socket shoulder joint, humerus, radius, ulna, even five phalanges. They also look like the fingers of a squirrel and the toes of a mole. This similarity—these finger bones hidden among cartilage in the fishy fins—is another reminder that dolphins' ancestors once lived on land.

Only incredibly long tail muscles could support such antics.

Stay Cool Week 12

After 12 weeks of pregnancy, the dolphin fetus is now a quarter of the way from conception to birth. We can see that it's male; over the last week the penis has become visible. Before birth, however, the penis will be enclosed within the body. It will be housed in the genital slit (see page 68), to avoid creating unnecessary drag on the overall streamlined shape that is essential to moving efficiently through water. This body style also diminishes the percentage of skin exposed to the external environment, reducing heat loss.

Like many evolutionary adaptations, the dolphin's penis pocket creates as many problems as it solves. Sperm can't survive the usual mammalian body temperature, which is why mammals from dogs to human beings carry the penis and testicles outside the body cavity. So over time the dolphin circulatory system evolved a way to draw heat away from the genital area. A network of highly efficient blood vessels radiate from the testicles to run just under the thinner skin of the tail fluke and flippers and dorsal fin. Arteries carrying warm blood from the heart are surrounded by veins transporting cooler blood back from these extremities.

Most mammals automatically release heat through the mother's abdominal wall to maintain the healthiest body temperature for the young being carried inside. Dolphins can't do this, however, because their insulating layer of blubber is too thick, so the fetus swims in its womb, and this circulatory efficiency also helps keep him cool.

The dolphin's circulatory system boasts other tricks, too. As further insurance, when a dolphin dives, its blood flows away from tissues that have developed a tolerance for lower oxygen levels and moves instead to supply areas that require more oxygen than usual to combat underwater stress—brain, heart, and lungs. Dolphins' muscles also contain a high percentage of myoglobin, an oxygen-binding protein that stores this essential element and helps prevent muscle fatigue.

Birds are also warm-blooded, and they have evolved their own ways to remain cool in hot weather. Some dig shallow depressions in sand or soil, settle down, and spread their wings to expose as much body space as possible to both cooler earth and moving air. Birds with wattles or combs also aim those into the breeze, because these body parts have blood vessels close to the surface to help lower the body temperature.

Consider yet another strange adaptation. Dolphins never evolved an ability to drink salt water; they must meet the universal mammalian need for water in some other way. Their kidneys can't desalinate ocean water, but they have evolved the ability to retain as much water as possible after distilling it from the fish that they eat. Although this dolphin will spend his entire life in the ocean, he will always be like the Ancient Mariner—surrounded by water, water everywhere, without a drop to drink.

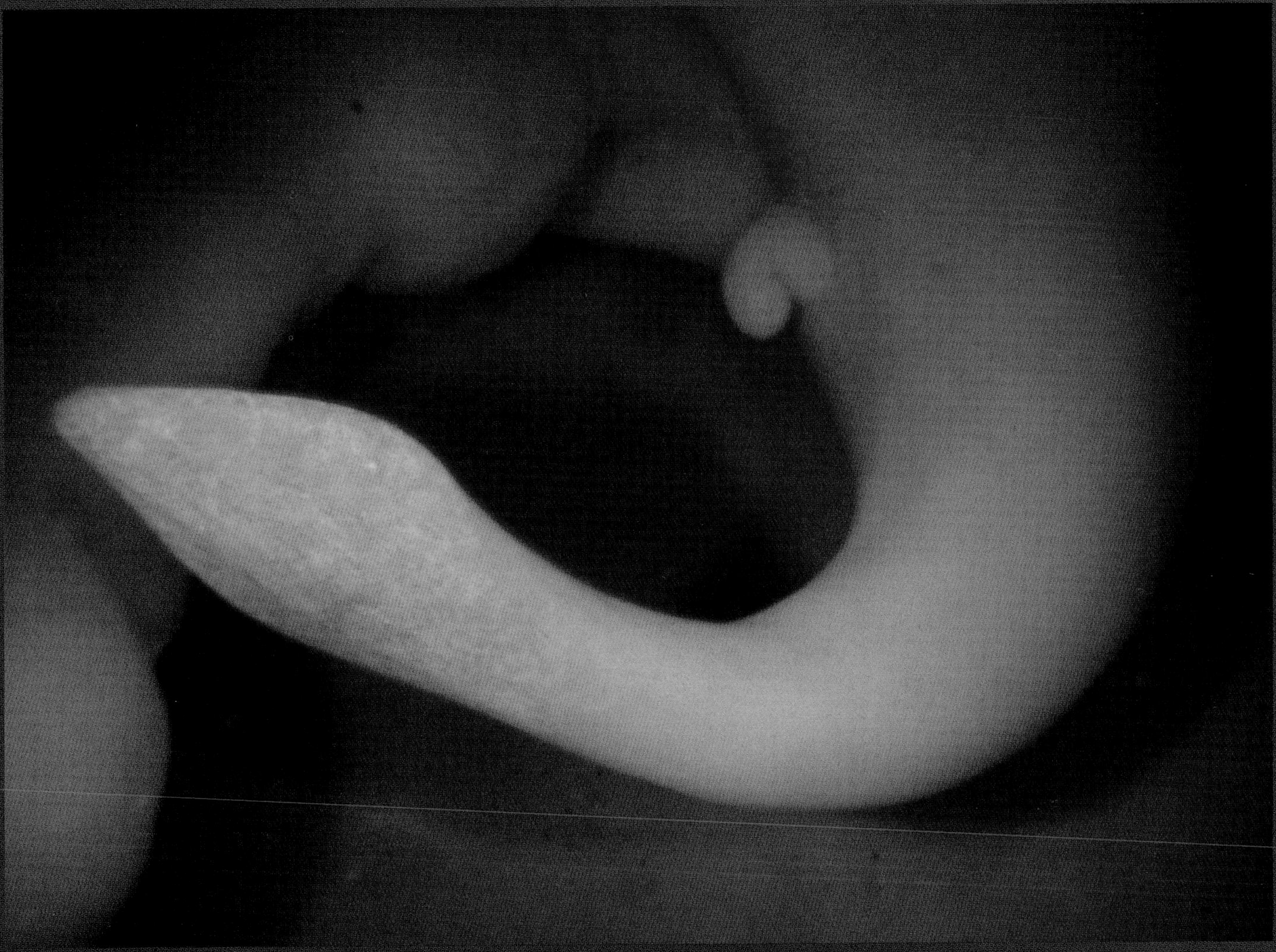

The penis, visible at the upper right, is still outside the body, but will move inside a pocket before birth.

The Shape of Things to Come

The limbs of animals have evolved into an impressive array of specialized tools for particular jobs. Just consider the similarities and differences between the limbs of those animals that we have been following in this book. A dolphin's flipper has more in common with a dog's foreleg than we might realize, just as it has less in common with a shark's fin than you would think.

"If we view a porpoise on the outside, there is nothing more . . . [like] a fish; if we look within, there is nothing less."

Edward Tyson, 17th century

Biologists are careful to distinguish between two kinds of physical similarities—homology and analogy. Homologous body parts are those that resemble each other structurally but now perform different functions, such as the forelimb of the golden retriever, bottlenose dolphin, Asian elephant, and human being. The distinct similarity of structure between the limbs of these animals—from finger bones to shoulder socket—indicates a shared ancestor. But subsequently each limb has evolved in response to the demands of its own environment.

Analogous structures, in contrast, evolved similar functions but from different ancestry and in response to the rigors of different habitats and lifestyles. The classic example of analogy is the animal wing. A luna moth, a California condor, a vampire bat—the ancestors of each evolved their own kinds of wings. But each variation still achieved its purpose: lifting the creature into the air, so that it might either chase or flee predators in a new dimension. The eye is another good example. Many creatures evolved their own version of an instrument that would detect and respond to the visual spectrum of electromagnetic energy, at whatever level of sophistication. The renowned evolutionary biologist Ernst Mayr once calculated that such organs had evolved at least 38 times.

As we have seen, many animals have changed environment completely over the eons—land animals marching into the sea, marine animals crawling onto shore, birds leaving the air behind and choosing land or sea instead. Over millions of years, the wings of penguins changed into flippers. All penguin species have the same shape of flipper: fatter and more rounded toward the front, and smaller at the back edge. This design reduces both drag and turbulence. Because of their common ancestry, penguin flippers are homologous to the wings of falcons and hummingbirds. And yet, in an interesting convergence, because of their similar function they are analogous to the fins of fish and sharks and of those other efficient swimmers in our story—dolphins.

Familiar-looking finger bones hidden inside the dolphin's flipper

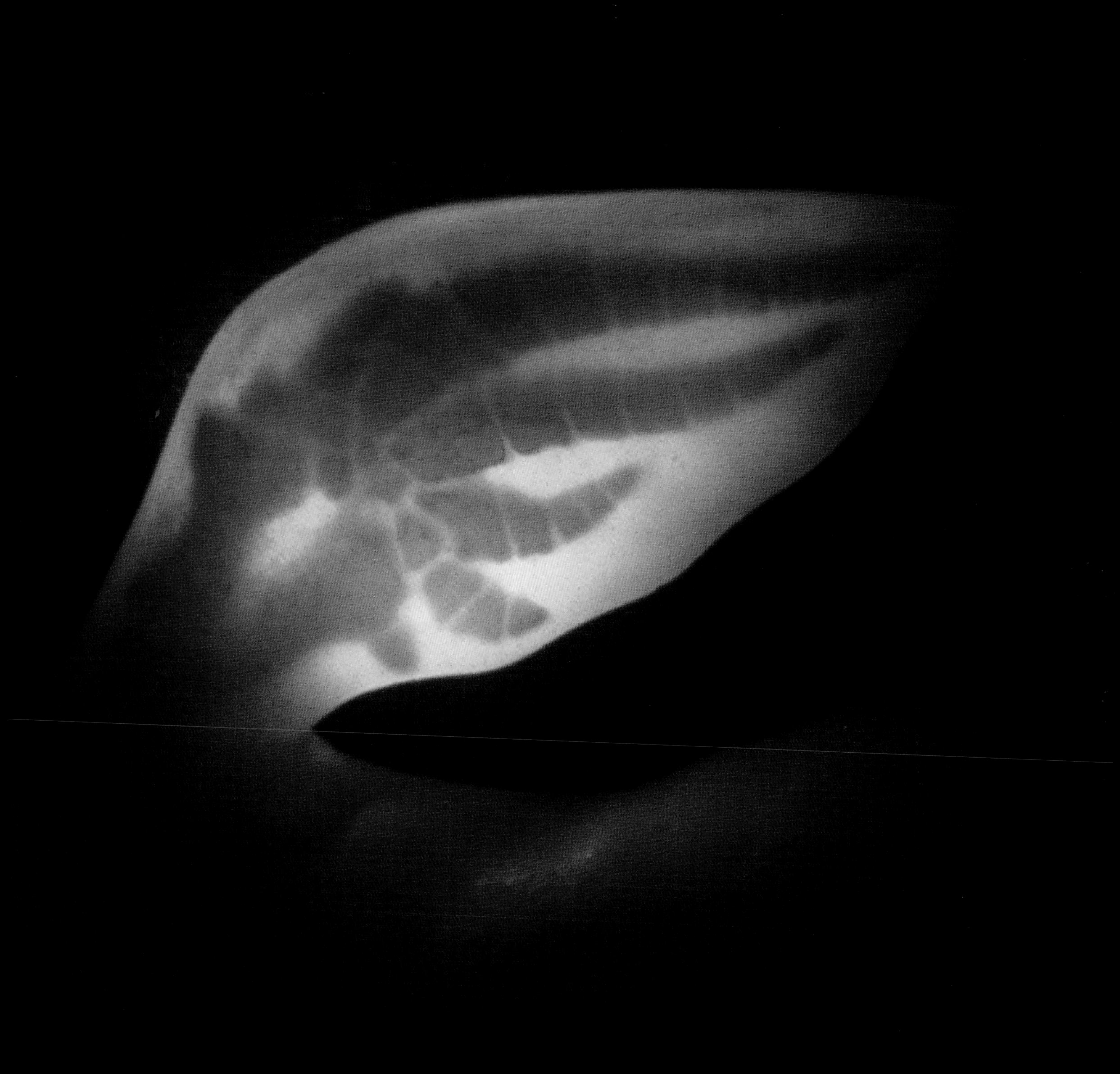

Blip

During millions of years of adapting to their wildly varied lives on Earth, animals have evolved countless ways to locate and pursue prey. Wolves can follow a scent trail accidentally left by a deer's feet days before; sharks smell blood and other substances in water at amazing distances. Owls hear the rustle of mice, and with ultraviolet vision kestrels spy the urine trails left by voles. Turkey vultures fly low above ground to detect the odor produced by early stages of decay in fresh carrion. And dolphins broadcast sound waves with their forehead and listen for the returning echo that bounces off fish.

Dolphins produce many different sounds, usually differentiated as whistles, clicks, and what the specialists call burst pulses—chirping or mewing noises that the animals produce in stressful situations. They don't actually speak, by our definition, because they lack vocal cords, but they have plenty of other ways of producing sounds. Those they make in air and those they make underwater require two different noisemakers. For the chatter and squeaks you overhear at Sea World, dolphins release air through the blowhole, the muscular flap of which they can control almost as precisely as we shape our lips to form different words.

Each dolphin also develops a signature blowhole whistle, at any time from one month to two full years after birth; apparently they aren't born with this talent but have to learn it. They're impressive mimics and sometimes imitate each other's particular whistle, as well as sounds they invent out of spontaneous play and in response to human nudging. In captivity, dolphins are known to call to people who walk by their pool.

But their most remarkable accomplishment with sound is echolocation—a built-in high-tech system that bounces sound waves off objects and deciphers an amazing amount of information from the resulting echoes. This sophisticated hunting system is what submarines imitate, in their clumsy way, when they use sonar to navigate underwater. The term "sonar" was coined during World War II as an acronym for "SOund NAvigation and Ranging," but animals have been using sound waves in this manner for millions of years. Donald Griffin, the American animal behaviorist who first documented this phenomenon in bats during World War II, dubbed it echolocation. To distinguish it from human technology, the animal version is also called biosonar.

Like its familiar squeaky chatter, a dolphin's navigational sonar clicks emerge not from its grinning mouth but from within the head above it. The sounds are produced so quickly, in series called click trains, that we hear them as a buzz; we can distinguish a single click only by recording and slowing a series. The dolphin produces these sounds as if inflating a balloon and letting the air squeal out through the knot.

Behind the blowhole lie three nasal sacs. After a dolphin surfaces and opens its blowhole to breathe, some of the air inflates these sacs, each of which is sealed with a plug that produces sound as air passes through it. These relatively faint sounds are amplified by projecting them through a waxy structure nicknamed, because

An MRI reveals the sonar-operating skull and blowhole.

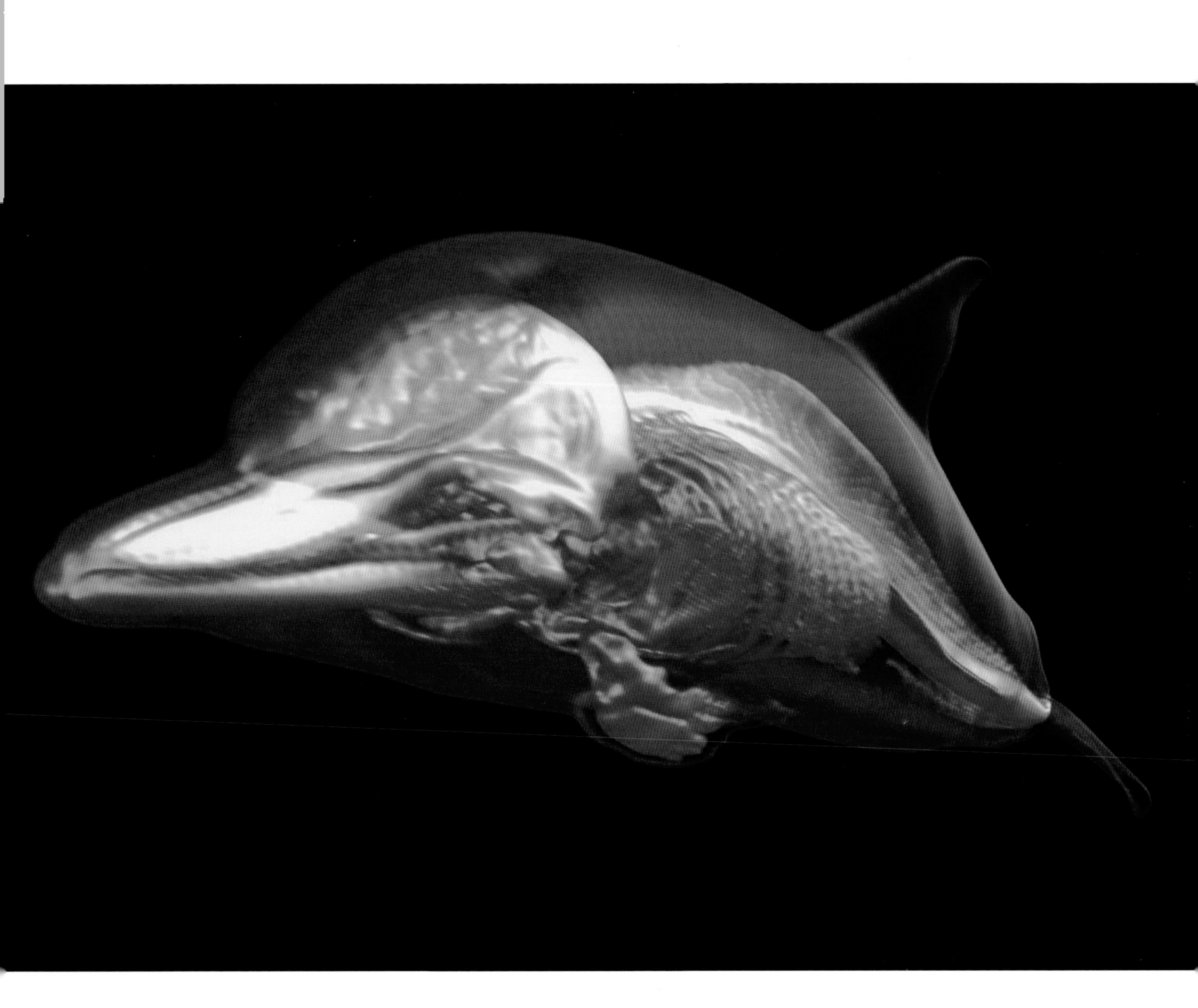

of its size and shape, the melon. In the adult, it will occupy much of the forehead, in front of the skull and blowhole. In the fetus, around the age of 13 weeks, the melon begins to appear as a curious bulge in the silhouette of the head.

At about 42 weeks of development, the fetus will grow tiny holes just behind the eyes and above the end of the jaw. They don't look like ears to us, but these orifices permit sound waves to enter the middle and inner ear, whose sensitive connections are similar to our own and those of other mammals. Sound enters the inner ear through the lower jaw. The bone there is thin and surrounded by fatty deposits that reach all the way to the tympani of the inner ear. Scientists report that covering the lower jaw with neoprene results in considerable loss of hearing.

Dolphins can detect sounds within a frequency range from 1 to 150 kHz—topping out much higher than the dog we found so impressive earlier, with its limit of 35 kHz (see page 42). The dolphin's brain is proportionately almost as large as our own, and its advanced auditory centers are one reason. It takes a lot of hardwired brainpower to process the signals coming from as many as 68,000 cochlear fibers feeding into the auditory nerve—more than twice as many as we have.

Like just about everything else in the natural world, echolocation is even more complicated than it sounds. Dolphins use lower frequency sounds, which travel farther, to detect distant objects. As they get closer to the object, they raise the sound frequency because, although high-frequency sounds don't travel as far, their up-close echoes convey more detailed information. This system is so precise it can detect an object the size of a Ping-Pong ball more than a hundred yards away. And a dolphin must not only receive echoes but also interpret them, answering numerous questions as quickly as possible: What is this object that bounced back my click? Precisely where is it? How far away is it? In which direction is it moving? How large is it? How dense?

This last question must be part of the information that dolphins process, because their sonar detects a unique kind of signal from the swim bladder of fish. This gas-filled sac helps fish maintain buoyancy. Although fish in general have a density similar to that of water itself, and thus register only slightly on echolocation mapping, their swim bladder reflects sound waves in a distinct way. Dolphins' echolocation system measures the direction and distance of blips from these inflated, low-density swim bladders. And our dolphin will be able to perform all these tasks while other dolphins around him are clicking and squeaking their own sonar system and deciphering the many returning echoes. Out of this cacophony, dolphins manage to navigate, hunt down food, and communicate with their fellows.

This Navy-trained dolphin's transponder tracks his location.

Handling the Pressure Week 24

Our dolphin is almost halfway through his year in his mother's womb; it has been 24 weeks since conception. Many changes have taken place. Perhaps the most obvious is that he is changing color from pink to gray. As an adult, he will be slate gray on the back and a paler tone underneath, ranging from pearly to pinkish-gray. This kind of coloring—looking dark like the sea from above, light like the sky from below—is called countershading. It helps dolphins fade into their environment. You can also observe it in shore birds such as plovers, which use the same optical illusion to become less visible in the open. When their darker upper part is well lit and their lighter underparts in shadow, their shape against sand tends to blur almost to invisibility. Dolphins demonstrate the aquatic version of this adaptation.

Everything about his prenatal life is changing. To find a better position within his cramped home, for example, he has curled into an odd position, bent in a U-shape, beak-to-tail, folded around the inside of the uterine wall. Oddly, he is bent sideways instead of forward. This curious posture creates folds in the skin that will remain visible for several weeks after birth. He can manage these contortions because five of his seven neck vertebrae never fuse together, which will always keep him much more flexible than most other marine mammals.

He has also grown a tail that we can now truly recognize as a dolphin's. By the time he's ready to leave the womb, his flattened horizontal tail flukes will spread to about 20 percent of his body length, which will probably be around nine or ten feet (and as an adult will be significantly larger than a female). Driven by the robust trunk muscles, the tail creates a powerful form of propulsion, even though it lacks both muscle and bone and consists entirely of dense, fibrous connective tissue. Unlike a fish's tail, which moves sideways, the dolphin's will move up and down. It will be strong enough to propel him out of the water in the spectacular leaps that we so admire, rising as high as 15 feet into the air. Scientists have proposed several theories about why dolphins leap so far—to herd fish toward other dolphins, to help remove parasites, to look for predators. Also, because dolphins are playful social animals, they may just enjoy cavorting.

"There is about as much educational benefit to be gained in studying dolphins in captivity as there would be studying mankind by only observing prisoners held in solitary confinement."

Jacques Cousteau

He will need more than propulsion, of course, to survive the challenges of life in the sea, so other adaptations are developing on schedule. One is a powerful and efficient circulatory system. As early as three weeks after conception, when the embryo was still only half an inch long, scientists could already detect the first hint of a heartbeat. Over the next five weeks or so, it developed more fully, reliably performing the rhythmic activity that will result in a resting heartbeat similar to our own—roughly 70 to 80 beats per minute. Dolphin blood has both a higher population of red blood cells and a greater concentration of hemoglobin within each cell than ours, which means that it is much more efficient at transporting oxygen. On average, this dolphin's heart will beat more than a billion times, pumping blood for, if it's lucky, 30 years of facing how the sea challenges a creature whose ancestors evolved on land.

Diving to any considerable depth places the kind of demands on dolphins that are familiar to scuba divers. A physiological disorder called "the bends" results from too much nitrogen quickly accumulating in the blood. Nitrogen makes up more than three-quarters of Earth's atmosphere; dolphins, like human beings, inhale a great deal of it with every breath. Human divers must breathe air that has been pressurized to match the water pressure around them. But the higher the water pressure, as divers go deeper, the more quickly nitrogen dissolves into our blood. As a consequence, if we try to return to the surface too quickly, nitrogen bubbles form in our blood and cause pain, sometimes paralysis, and occasionally even death.

So we move slowly and let the nitrogen work its way out of our system. But we aren't down there chasing lunch. Dolphins are—and they're doing so without scuba gear. Often they must move quickly, both diving and surfacing. They have evolved several ways of responding to the pressures of their environment, each of which is developing in this fetus. For one thing, he will be able to hold his breath for as long as seven minutes. On most dives, he doesn't breathe at all and thus avoids polluting his blood with nitrogen. Dolphins have been known to dive as deep as 3,000 feet, and he will have no trouble at all in reaching 500 feet. He will also automatically lower his heart rate to only a dozen beats a minute, which greatly reduces blood flow to every part of the body except the brain and heart.

Without equipment, humans can dive to no more than 100 feet and remain at that depth for a maximum of three minutes. If we dived deeper, the water pressure would crush our ribs, which form a rigid cage for the lungs. The rib cage of dolphins, in contrast, has several ribs that are anchored on only one end and jointed to enable them to fold. Before diving, our dolphin will take a deep breath at the surface and lift his tail out of the water and turn upside down to give gravity some help when he plunges. As he dives deeper and deeper, his lungs actually collapse, and the jointed ribs facilitate this strange adaptation.

Double Vision

Don't ever try to sneak up on a dolphin. You'll be wasting your time. Their eyes, situated on each side of the head rather than near the front, provide the widest possible field of vision—forward, backward, up, down, for almost a full 360 degrees. Now, at 28 weeks, the fetus's eyes are already opening and closing in the womb. Like a chameleon, he will even be able to move each eye independently, further increasing his already formidable optical equipment.

Other adaptations refine dolphin's underwater vision. At the corner of each eye is a gland that secretes an oily mucus that performs several useful functions every time the dolphin's blink spreads it across the cornea. It washes away debris, of course, and ocean water is full of floating organic and inorganic debris. But it also lubricates the eye and helps guard against infection. It may even help streamline the eye so that water flows over it as efficiently as over the rest of the animal's elegant shape.

Surprisingly, dolphins can see as well above water as below the surface. Eyes that evolved in terrestrial animals to enable them to see through air usually can't perform anywhere near as well underwater. We ourselves, for example, wear dive masks in part to reduce distortion. But cetaceans can see well in both environments, despite having adapted in so many ways to the sea. We might have guessed at this versatility, considering how quickly dolphins react to people on shore and how easily they catch fish thrown to them, but experiments and dissections have now confirmed it. The eyes of dolphins, like those of fish, are spherical, enabling them to focus light on the retina despite underwater distortion. Strong ocular muscles contract and relax to adjust the shape of the lens—more rounded for seeing in water, flatter when necessary for seeing in air.

Convergent evolution results in the kinds of analogous structures discussed on page 90. For example, even though they evolved along completely different paths, human eyes resemble those of cephalopods (octopuses, squid, and their kin). Never mind that their eyes have an outward-facing retina and ours faces inward. Forget that squid eyes develop from nerve tissue and human eyes from skin tissue. Amazingly, both feature an iris, a lens, and lids.

This double vision provides dolphins with predatory opportunities not available to exclusively aquatic animals. Biologists call one such ploy "strand feeding." Working together, dolphins rush through shallow water, creating a bow wave that flings fish onto the shore. Then, with their versatile lenses automatically changing to enable them to see above the surface, the dolphins scoop up fish as they flop around in the mud, out of their element and doomed to serve as a dolphin's lunch.

Dolphins can see well both underwater and above the surface.

Skin Deep Week 36

By 36 weeks, the fetus has grown a half dozen whiskers on both sides of the snout, each hair about a quarter of an inch long. Shortly after he emerges from the womb, water pressure will make the hairs fall out. They won't grow back, but the follicular pits, like pores, will remain open and visible on the rostrum; you can easily glimpse them at an aquarium. Even without the hairs that formed them, these follicles may provide sensory data, perhaps about water currents.

At this period of development, all the fetus's organs are in place and recognizable. His skin is officially dolphin-colored, having long since lost its early pink hue. Some species of dolphins have more patterning on the skin, such as spots, and those would be in place by now too. But the bottlenose has only elegant shades of gray, like blended charcoal.

The dolphin's smooth skin has no sweat glands. Dogs, as we have learned, sweat only through their feet, and they must pant to reduce heat and maintain the internal ecosystem (see page 44). Marine creatures have different needs; dolphins don't have to worry about reducing heat. As every swimmer knows, it is much more difficult to stay warm in water than in air of the same temperature, because the thermal conductivity of water is about 25 times higher.

Even though dolphins frequent tropical and temperate waters and stay away from the far north and south, their primary concern is maintaining body heat, not losing it. They achieve this goal with a high metabolic rate and a thick layer of insulating blubber. When it gets wet, fur loses much of its value as insulation, and all marine mammals have replaced it to some extent with blubber. Cetaceans long ago lost their fur, although some semi-aquatic species have not, as you can see from the photograph of sea lions on page 13.

Blubber serves as both thermal blanket and suit of armor. If a dolphin is attacked by an orca or a tiger shark, the predator's teeth may well sink no deeper than the flak-vest thickness of blubber instead of slashing vital organs—thus giving the dolphin a crucial moment in which to wriggle free and escape. Many dolphins wear attack scars, but dolphin mortality as a consequence of shark attack appears to be relatively low.

This dolphin's skin will provide other services as well. It helps hydrate bodily systems by serving as an osmotic membrane, letting in water while filtering out salt. And this remarkable skin performs another noteworthy trick: It flakes off constantly, entirely replacing itself every couple of hours, staying smooth and flexible and streamlined.

The skin is growing insulating layers.

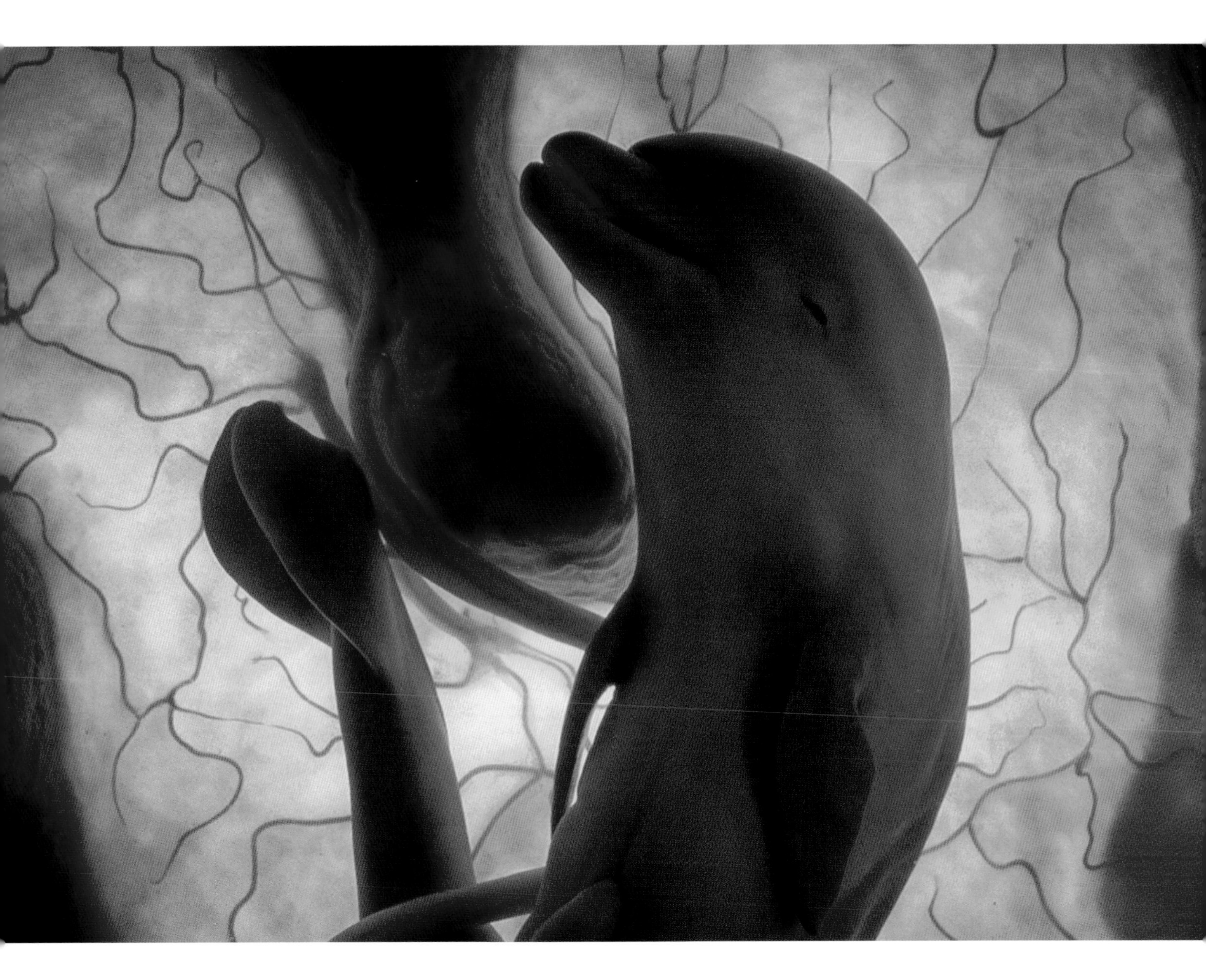

A Year in the Life

For a year, during a full cycle of the seasons, a full circuit of the Earth around the sun, the dolphin has been growing inside his mother's womb. The other animals we've glimpsed during this time have long been out in the world. Even a human mother who had become pregnant at the same time as this dolphin would now be caring for a three-month-old baby who would already be suckling and grasping toys and smiling at her parents.

The joey peering out from her cozy cradle

Eleven months ago, after only four weeks in the womb, the kangaroo joey was born in Australia and crawled as quickly as her oversize front legs could carry her into her mother's pouch. For six months she stayed in the pouch, hiding away from light and danger. Gradually she began to crowd her surrogate womb and began poking her head out to peer at the world. She was like a bird looking out from a nest—only even more protected. Then one day, about four months ago, her mother relaxed her pouch muscles and tilted forward and tipped the joey out to make her way in the world on her own big feet. Like most other mammals, she scurried for her mother's protection at every alarm, continuing to suckle as she learned to graze. Now, 11 months later, she is able to survive on her own and no longer returns to the pouch, but she may suckle for a few more weeks.

A month after the kangaroo was born the size of a cherry, while the dolphin was only one-sixth of the way toward birth, the penguin chick tapped its way out of the egg and into a life on the Antarctic ice and its almost equally icy water. Rumpled, its feathers four times as dense as those of other birds, the emperor penguin faced challenges as dramatic as the kangaroo's desert habitat or the dolphin's underwater realm. After remaining within a large crèche of other hatchlings, it left the nesting area about six months ago—still not fully grown and still dressed in downy feathers instead of the adult tuxedo. It has since moulted and reached adulthood, but will not breed until about the age of five.

About the time the penguin hatched, ten months ago, the golden retriever brought her pups into the world. Four months ago, at the age of six months, the pups began to reach sexual maturity. By now they could be having pups of their own, continuing the primordial cycle of birth, building the future.

A newly hatched chick keeping warm on its parent's feet

Tail First Birth

We have reached our second destination on the calendar of our animal pregnancies. It is week 52 and our bottlenose dolphin is about to give birth. It has taken a year for the next generation of dolphins to grow from a microscopic egg and sperm to an embryo, a fetus, and finally a baby dolphin crowding his mother's womb as if eager to get out and into the sea. He is three feet long already and weighs about 30 pounds. Not surprisingly, this descendant of land animals faces extraordinary challenges in being born underwater. By comparison, the golden retriever pups had it easy.

> Our newborn's dorsal fin is pristine and unscarred. As they grow older, most dolphins, especially in the wild, wind up with a dorsal fin nicked by fights with other dolphins and with larger predators such as sharks or even scarred by encounters with propellers.

He is moving down the birth canal, positioning tail first, so that he won't inhale water during birth. To protect his mother from injury during delivery, his flippers are folded against his small, sleek body, his tail flukes likewise limp and out of the way. Over the next few weeks they will grow firmer. At first, though, the dolphin will bear stripes and creases on his pearly gray skin, showing where his appendages were folded and stored for his first journey.

He is born tail first. His mother may swim around for some time with his tail visibly protruding from her lower body, until her uterine contractions push him farther. As blood clouds the water around his mother, the rest of the body emerges, then finally his head. His umbilical cord—the physical connection between generations—snaps. As of this moment, although his mother will nurse him for at least two years and probably longer, he is an independent creature.

Unlike the retriever pups or the kangaroo joey, the dolphin is born ready to face his undersea world. His eyes are open, his ears working, his muscular coordination developed enough from the first to let him swim along with no more than an occasional protective nudge from his mother. He may be around for 20 or 30 years, perhaps even the upper limit of 35. In general, females live longer; records show that some have passed the half-century mark. Like redwood trees, dolphins show their age in rings—in their case, from the layers inside their teeth (see page 138 for similar rings in sharks' scales).

Males and females mature sexually at different ages. For a particular individual, the time could be anywhere within a range of several years. It will be three or four years from now, at least, before this male reaches maturity and begins to play his own role in the future of his species.

A dolphin giving birth while swimming

following pages: **Nudging the newborn toward the surface for air**

Life Between Two Worlds

Often a dolphin mother about to give birth will be accompanied by another adult female—a perfect example of allomaternal behavior (see page 66 and 124). It is difficult to resist thinking that she is midwifing the mother, and that her attentive care later, when the mother is away, constitutes babysitting. These attending females have long since been nicknamed "aunties."

In the wild the baby instinctively swims toward light and air, often with close help from his mother. In captivity, with birth taking place in a large water tank, mothers sometimes guide the calf all the way to the surface, keeping an eye on its direction. Zoologists speculate that they do so to ensure that their calf won't bump against the sides of the tank during its first and most urgent swim.

Soon the newborn dolphin will find his mother's nipples, which are on both sides of the genital slit. At first he will suckle briefly every 20 minutes or so, 24 hours a day. In time he will nurse less often but for longer periods. Dolphin milk is concentrated and highly nutritious—7 percent protein compared with our 1 percent, 33 percent fat compared with 4 percent in human beings. He will grow quickly. A creature that is 80 percent muscle needs a lot of fuel. He will return to it for some time after he learns to fish for himself, apparently seeking not only physical but emotional nourishment—renewing the bond between mother and calf.

But nursing must wait until after the calf meets a more urgent need. As he emerges from his mother's uterus, he moves quickly. Within seconds after birth, he is at the surface, taking his first deep breath of air on his own—his body beautifully adapted to life in the water but still demanding the air that his ancestors breathed on land. This life in two worlds, this connection with both land and sea, must be part of what draws us to dolphins. Like birds, they inhabit a sphere that seems alien to us but that irresistibly lures our imagination. In our minds, the sea has been home not only to whales and squid and fish, but also to sirens and mermaids and sea monsters. To an air-breathing land-based animal such as ourselves, the vast worldwide ocean seems fathomless and temperamental. But it looks a bit friendlier as this newborn dolphin takes his first deep breath and dives back underwater with his mother.

The newborn suckles as he swims with his mother.

Parasitic Wasp

Lemon Shark

Part 3 Extreme

Asian Elephant

Giant

Standing near an elephant is a humbling experience. When a beast the size of a truck—an animal that could crush a human being with a single misstep—behaves affectionately toward its bumbling offspring, it does more than inspire a chorus of "Awwwww" from zoogoers. It reminds us that *Homo sapiens* is only one little branch of a sprawling family tree of sociable, warm-blooded creatures. Like dogs and dolphins and humans, elephants expend a great deal of effort to bring the next generation into this world—and their unique way of going about it tells us a lot about their history.

Just as the dog's Latin name, *familiaris,* reflects our view of the species, so does the Asian elephant's. *Maximus,* scientists have dubbed this species—*Elephas maximus,* the giant, the extreme. Although the Asian elephant can reach a height of more than nine feet and a weight of over five tons, it still measures a bit smaller than its African cousins. Pliny the Elder, who spoke so highly of dolphins, devoted many pages of his nature encyclopedia to the elephant, declaring it the animal "closest to man as regards intelligence." The notoriously gullible Roman also insisted that elephants are honest, fair, and respectful of religion, and claimed that they mate in secret because of their modesty.

The easiest way to tell apart living elephant species is by their ears. The African's are much larger, covering the entire side of the head, and they helpfully resemble the shape of their native continent. If you compare the species at a zoo, you will also notice that the Asian elephant has a more arched back and a bulging, rather than flat, forehead. It handles plants differently, too. At the end of the trunk it has only one "finger"—the semi-prehensile appendage that helps in grasping—as opposed to its African cousin's two.

No one knows precisely when either branch of the family was first domesticated, but Asian elephants are mentioned in Sanskrit epics from more than 3,000 years ago. They were weapons of war in India until the invention of the cannon, and in civilian life they continued to serve as logging trucks. In cities such as Bangkok you can still glimpse them on the streets. Poaching and habitat loss have greatly reduced their range in the wild; nowadays you can find them only in parts of India, Thailand, Sri Lanka, Indochina, Indonesia, and Bangladesh, where they follow migration routes in response to monsoon seasons. Most Westerners, however, see elephants only in zoos and circuses, ruminating over food or elegantly caparisoned for a parade.

Our Asian elephant's pregnancy will last 10 months longer than the dolphin's, and 20 months longer than the dog's. Not surprisingly, the largest land animal also has the longest gestation period. Elephants demand superlatives.

The famous spring elephant march in Trichur, India

***previous pages*: A herd of elephants blocking a trail in India**

THE GRE
LEPHANT
A R C
51

Call of the Wild

You can tell male and female elephants apart by their tusks. Actually just enlarged upper molars that don't stop growing, tusks fall out as they age and are replaced by other molars moving forward and taking their place. The cow's tusks, called tushes, aren't visible beyond the lips, but the bull's curve outward and upward in the impressive array that, tragically, tempts ivory poachers.

In the wild, although they forage over a wide area, the elephants seldom stray far from river or lake. Their gargantuan bodies demand anywhere from 20 to 40 gallons of water daily to accompany their insatiable appetite for just about any edible part of a plant—grasses, leaves, bark, fruit, vegetables, roots. Naturally their appetite is as extreme as their size. They eat at least 165 pounds of food a day and sometimes twice as much, although not all of it is well digested, and they leave behind a predictably elephantine amount of waste matter.

Male and female Asian elephants lead separate lives most of the time. The cows tend to herd together with their young and their female kin in groups of 20 or more. The young nurse for about a year and a half and then remain with the family until around the age of 14. Bulls are more solitary. They wander away from their natal herd at this age to test themselves against other bulls and establish dominance hierarchies, remaining alone or in small groups of other lonely adolescent males.

Mostly the adult sexes come together only when the female enters estrus, which happens roughly every four months and lasts for two or three days each time. This cycle means that she is fertile three times a year but only for a brief period each time; cows are receptive to mating only on the first day or so of the cycle. Following her hormones' command, the cow begins to broadcast a low-frequency mating call that the males can hear as far as three miles or more.

Elephants don't win the award for longest bouts of mating. Voyeuristic zoologists have recorded lions mating every 25 minutes for up to 300 times. And mouse-size Australian marsupials of the *Antechinus* genus literally mate until death do they part—in trees, with more than a dozen partners over 12 hours or more. Males soon die, some while mating, and the females raise the young.

These calls are mostly infrasonic, below the range of human hearing, but some frequencies are audible to us. Just as dolphins employ lower pitched biosonar when they want a signal to travel farther (see page 94), so do elephants use very low pitches to convey their message without deflecting too much off their leafy, hilly terrain. Asian elephants engage in calls throughout the year, with most of them produced between groups of cows. Such broadcasts help coordinate complex societies over wide and constantly changing areas. But the female's mating call is unique. Hearing this rumbling sound, the bulls launch into the swagger and brawl that will produce a victor that then woos the cow.

Unlike dogs and dolphins, elephants can take their time mating.

The Great Race Month 1

Many of nature's mating strategies demand elaborate rituals and acrobatics. Elephant sex is a case in point. Over three days or so, when not merely hanging out together and courting, the elephants will mate as many as five times daily. Not once, however, does the male's curvy three-foot-long penis actually penetrate the female's vagina. Instead it sprays at the vaginal opening a jet of sperm—more than eight ounces, or at least a hundred times the amount produced by a man. The vagina itself is a barrier as much

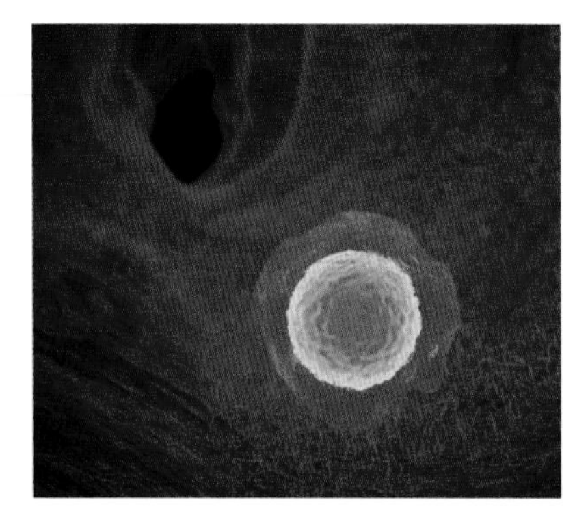

A ball of cells in the uterus

as a gateway; its many longitudinal folds permit only a tiny opening, less than half an inch across. In this haphazard system, a third of the bull's sperm don't even make it inside the vagina.

And the surviving sperm still face a triathlon of other challenges. After the elephants' long weekend of hit-and-miss mating, the bull wanders away to seek another female with whom to repeat his curious performance. He leaves the cow to her biochemistry.

Already she is responding to intercourse in a variety of ways. First, her vigilant immune system, which serves as an internal security network, perceives the sperm as foreign invaders. Leukocytes—white blood cells—constantly patrol the body to protect it against intruders, and they attack the sperm with every chemical weapon in their arsenal. Many more sperm fall in this way.

There are plenty of opportunities for the white blood cells to attack, because the sperm have to swim six and a half feet from the vagina's opening to the waiting egg. Human sperm, in contrast, must swim an average of only three inches. The elephant sperm's highly competitive swim meet is even more dramatic than the race-against-the-clock climb of the newborn kangaroo joey up to her mother's pouch. Survivors enter the uterus, where its muscular walls, lined with tiny waving hairs, helpfully speed them on their way.

Finally the sperm reach the oviduct. Its winding, narrow passages, which lead to the ovary, trap many more sperm. By the time they reach the egg itself, only 10,000 of the bull's 5 billion sperm are still in the game. It has been 12 hours since mating. Although laggards never make it, the race is not to the swift, for the first sperm to reach the egg is not necessarily the one that will fertilize it. The egg is guarded by a protective layer that only the right mix of proteins can penetrate. Eventually a single sperm gets through. As the egg engulfs it, the merger triggers a chemical change in the protective shroud that prevents other sperm from entering.

White blood cells perceive sperm as intruders and attack.

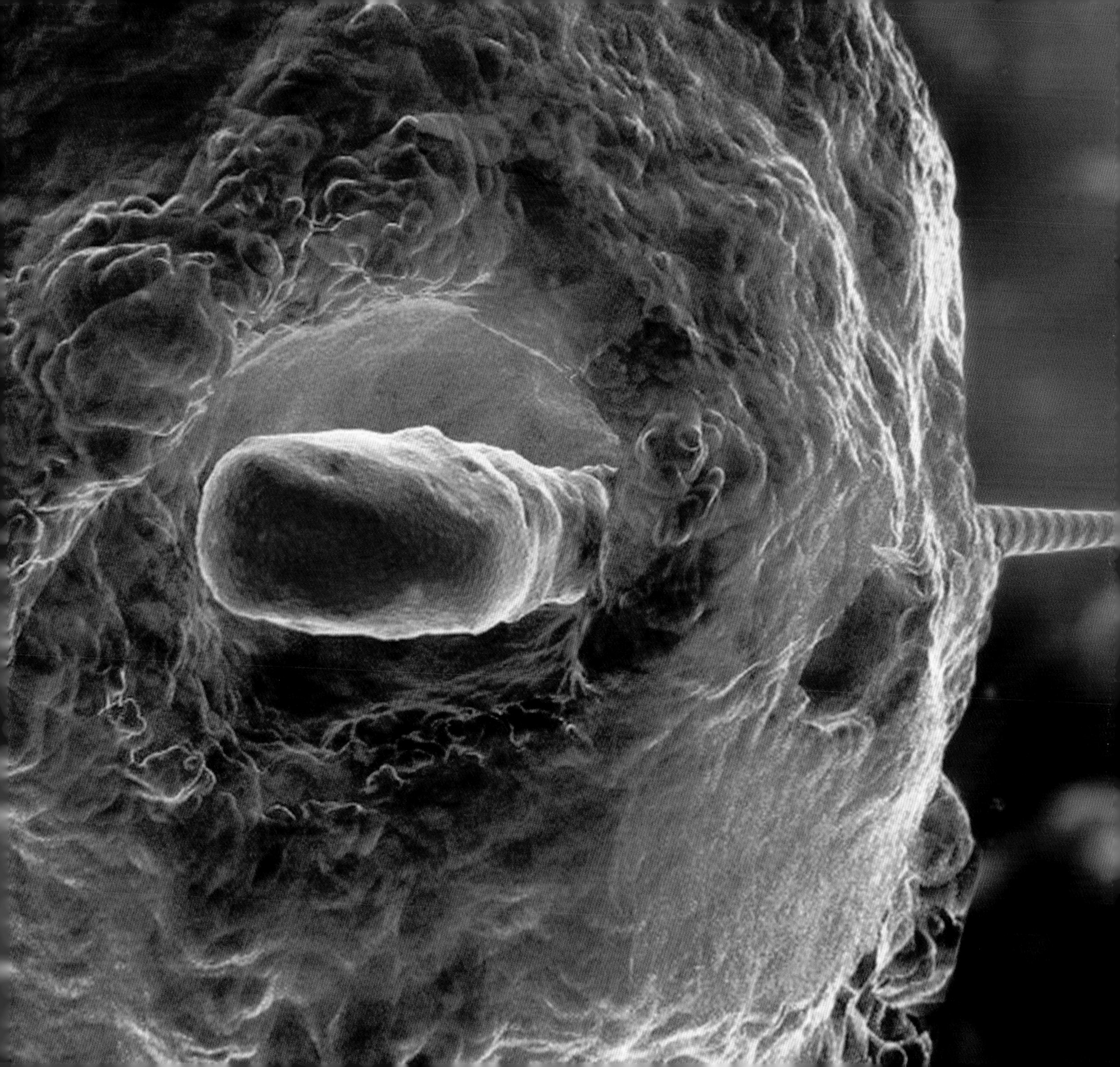

Everything

The sperm's chromosomes merge with the egg's, uniting the elephant's genetic heritage of father and mother in a zygote weighing no more than a millionth of an ounce. The entire blueprint for growing a giant animal lies hidden in the DNA coded into this microscopic egg.

When we think of the word "egg," most of us envision the egg of a chicken. For that matter, 21st century urbanites are so removed from the sources of our food that we might imagine a carton of eggs in a refrigerator rather than a clucking hen perched atop a single egg. Because of this familiar mental image, we associate eggs with birds but not with other animals. Yet all vertebrates and most other creatures begin life as a single fertilized egg cell—human and dog, dolphin and elephant, penguin and kangaroo, as well as the shark and wasp that we are about to start following. *"Ex ovo omnia,"* wrote the 17th-century English physician William Harvey. "Everything comes from an egg."

One primary difference between eggs is how much nutrition they carry, which is a rough index of how long they need to support an embryo. Consider the animals we have been following. The dog and dolphin grew from an egg firmly attached to a uterine wall, as the elephant is doing now. A mammal's egg doesn't need to be large or well stocked with long-term nutrition. Its growing embryo soon receives nourishment directly from the mother's body, via the umbilical cord and placenta.

The egg of the kangaroo, by contrast, floats in the womb instead of developing an umbilical connection, like a space capsule with no link to the mother ship, and therefore required to survive with whatever it carries on board. Hence a kangaroo's always premature birth, to be followed quickly by a scurry to the pouch and intensive nursing with nutrient-rich milk to make up for its lack of development while inside the egg.

A shell, either hard or rubbery, surrounds and guards eggs laid out of water—those of birds and snakes and lizards. Amphibians, even those that live primarily on land, lay their gelatinous masses of eggs in water, because otherwise they would dry out and die. Most female fish lay unfertilized eggs that are then fertilized externally by the male. These then develop on their own without further contributions or attention from either parent. Because this reproductive method requires considerably less investment from the parents, some fish, such as cod, lay millions of eggs in a single spawning. This adaptation is in stark contrast to the partnership between male and female penguins to support a single egg for months, or to the elephant's labor-intensive 22 months to produce one calf.

Because the egg is a single cell, we can say that the largest known cell is the egg of an ostrich, weighing in at more than three pounds. The West Indian vervain hummingbird has the smallest known bird egg, less than half an inch long and weighing 1/100th of an ounce. Insect eggs are, of course, much smaller. Tachinidae flies lay microscopic eggs, as small as 0.02 millimeter.

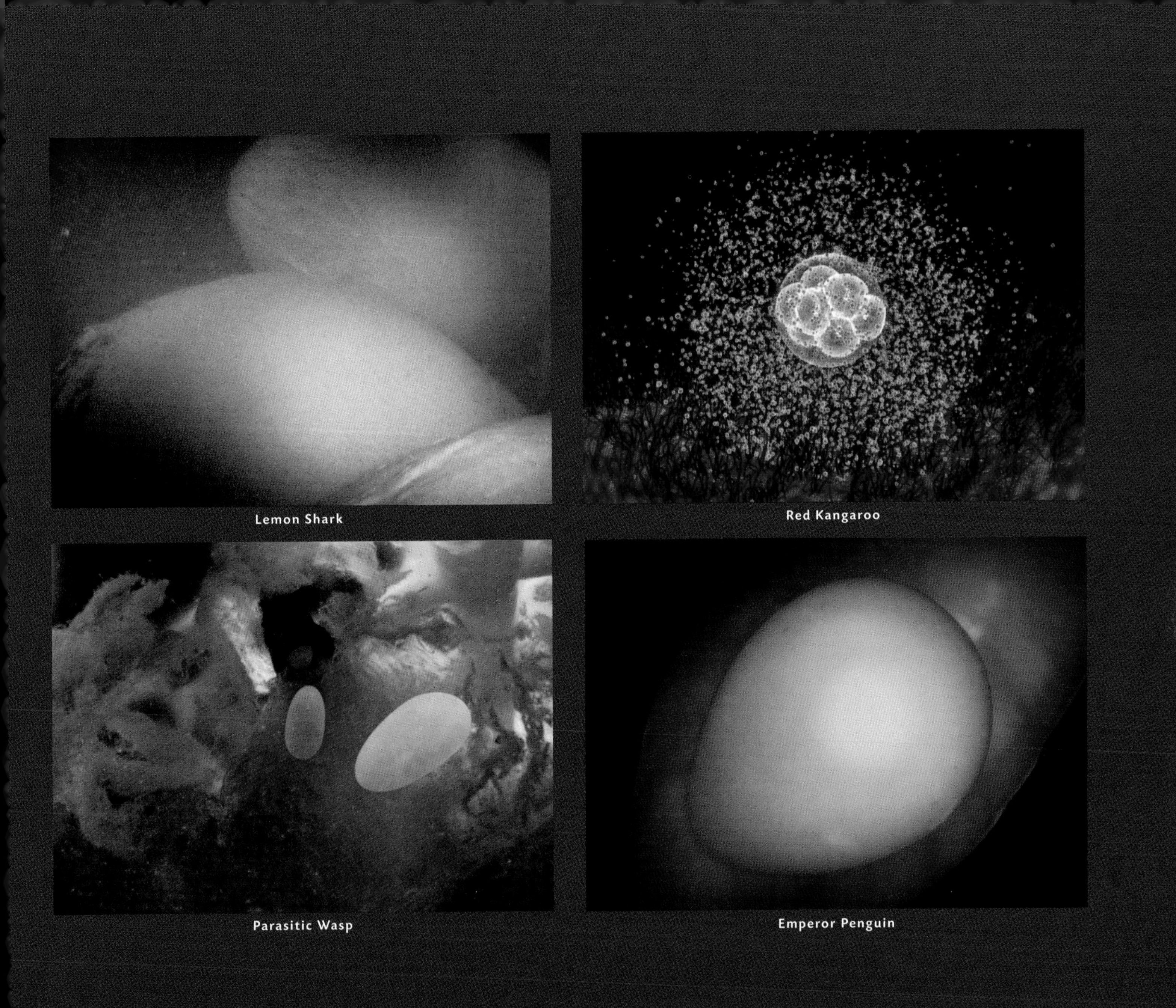

Different animals begin life in surprisingly similar ways.

Family Resemblance Month 2

For two months, the elephant embryo has been moving through the phases that we witnessed earlier in dog and dolphin. Soon after fertilization, it chemically signaled the mother to produce progesterone, which stopped estrus cycles and began to prepare the uterus. Cells divided into a blastocyst. This elephant's uterus bears scars from the mother's previous births, so the blastocyst had to find an untouched area to implant itself in the wall of the womb. Gradually it united with the mother's circulatory system and began to form the placenta—the lifeline between bodies.

In its early stages, the embryo still resembles that of other vertebrates. To a casual observer, it might be a dog, a dolphin, or even a human being. Scientists point out that, during this period of intrauterine development, an embryo resembles the primordial ancestor of its group. In the case of vertebrates, this forerunner seems to have been a relatively simple, fishlike creature that lived in the seas of the explosively innovative Cambrian period, half a billion years ago.

Even mammals can trace their family tree back perhaps as far as 200 million years, to the early Jurassic. Rich fossil evidence indicates that mammals have descended from a shrewlike creature that scurried around the feet of dinosaurs. Most of us would casually identify mammals as warm-blooded animals that have hair. Specialists, on the other hand, would probably start with the key point that all living mammals have sweat glands. Mammary glands, which produce the many kinds of milk that feed young dogs and elephants, are modified sweat glands.

Mammals also have a four-chambered heart, as well as a unique part of the brain called the neocortex. But these points matter only to scientists. To most of us, mammals are warm blooded and bear live young that nurse until they are capable of eating other foods. Some other creatures give birth to live young, so nursing afterward is part of the definition. The entire class of Mammalia gets its name from this unique way of feeding young. Eighteenth-century Swedish scientist Carl Linné (Latinized as Carolus Linnaeus) named mammals for their milk glands.

No other group of animals produces its own food for its offspring. This striking evolutionary development is one reason mammals have proved so versatile around the globe. There are fewer species of mammals than of most other groups of animals; there are twice as many kinds of birds and exponentially more insects. But many of the 4,300-plus mammals are record holders. Fastest animal? Cheetah. Largest? Baleen whales. "Taken all in all," writes biologist Colin Tudge, "mammals are by far the most various of vertebrates, in body form and in way of life."

The elephant still resembles other vertebrate embryos.

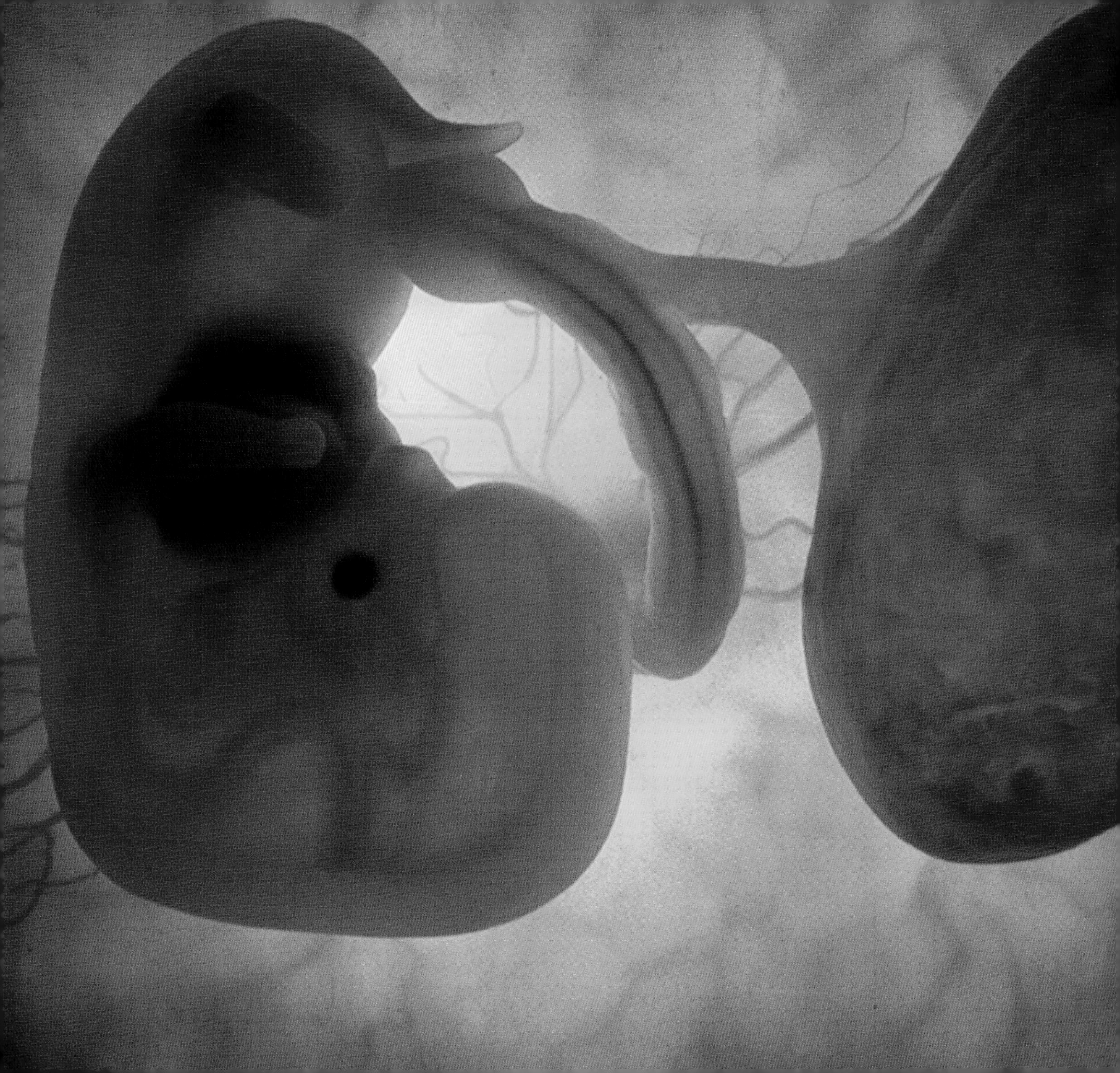

Budding Months 2 and 3

Until now, between two and three months after conception, the elephant embryo has resembled the early stages of other placental mammals. The embryo of a pig or chimpanzee would look much like this. For that matter, at two months after conception, the elephant's embryo looks very much the way a human one does at the age of about a month—an undifferentiated curl just beginning to shape itself. Scientists have identified many of the genes involved in forming patterns of body development; they turn out to be remarkably similar across a spectrum of animals.

But now the elephant embryo begins to grow tiny limb buds and the head starts to look more like a head and less like a lump. About the size of a dime—less than an inch long, weighing less than half an ounce—the embryo is connected to the wall of the uterus because the placenta is forming. The yolk sac can be distinguished. Then comes a faint heartbeat. By halfway through the second month, the embryo delineates a trunk shape and the elephant is entering a phase that we can more easily recognize as its species' own unique shape.

Scientists call a developing placental mammal an embryo from the time the zygote first starts dividing until it becomes a fetus. The problem with this definition is that there is no precise moment at which an embryo officially metamorphoses into a fetus—even though in human development we have defined this milestone as occurring eight weeks after conception. The rule of thumb is that an embryo may be called a fetus from the time that its major organs have formed. For nonspecialists, it seems that the key point is when we can begin to recognize its features as related to those of its adult form.

Like dolphins, elephants have been observed "babysitting" the young of other females (see page 66 and 108). This allomaternal behavior, as scientists call it, knits together the community and provides a safer environment for all of a herd's offspring.

We experience many such milestones during the growth and development of our own offspring. Recognition of the embryo's humanity is emotionally more compelling after its regions differentiate and predict their upcoming shape. This is the point at which future parents peer at ultrasound images of their embryo and exclaim over its miniature fingers. Resemblance is a powerful attraction. If we were forced to care for our offspring before they physically resembled their elders, we would have a much harder time developing an emotional attachment or valuing the offspring as one of us. Throughout history, this natural connection between appearance and emotional attachment has been tested by rare mutant forms—extra limbs, a hairy body—that societies have all too often designated as monstrous. Fortunately the elephant won't have this problem. Her offspring is developing normally. By the time she sees him, he will be unquestionably one of her group, undeniably her own.

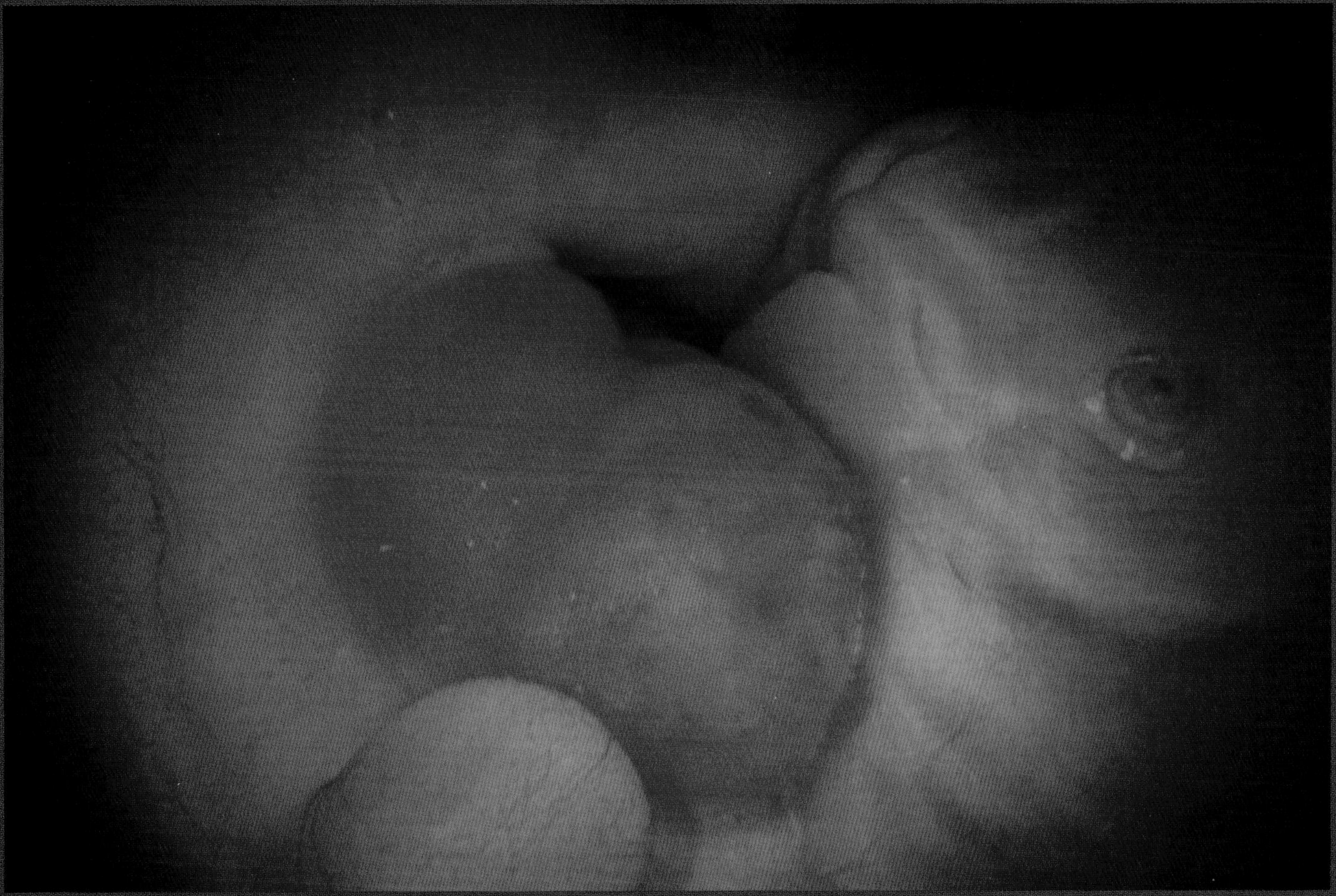

As limb buds and trunk bud sprout, the embryo begins to look like an elephant.

Strongheart Months 2 and 3

One of the most popular stars of the silent-film era was a German shepherd named Strongheart, but an elephant deserves this name even more. The faint heartbeat that begins to appear at this time predicts one of the most impressive characteristics of the adult elephant's body—its heroically powerful heart. This protoheart that barely exists at the moment, animated by a negligible rhythmic pulse, will grow into a gigantic muscle weighing over 50 pounds—more than half the size of Strongheart himself.

"Night is like the heart of an elephant."

Richard Sansom

A mouse's heart can beat 500 times per minute, while your own averages 70 to 75 beats during the same period. But when this elephant reaches adulthood, its heart will beat only 30 times a minute, once every couple of seconds. The same sort of ratio shows up in birds, which are also warm blooded. The heart of a tiny ruby-throated hummingbird may beat at a rate of 1,200 times per minute to power its blurring wings while it sips honeysuckle nectar. Yet a wandering albatross soaring on ocean winds will require less than a tenth as many beats per minute to keep its 11-foot wingspan airborne.

Although it may work in slow motion compared with many of its fellow animals, with every beat the elephant's heart will pump almost a thousand pints of blood throughout its giant body. The human body, by contrast, has no more than nine to twelve pints. A pint weighs about a pound. Therefore, this embryo's mother's heart—which grew from a start as modest as her offspring's—is pumping an amount of blood equal to about a tenth of her weight. And it can do so for as long as 80 years.

The embryo's brain has also barely begun to differentiate. Now it looks like a water bubble, but in adulthood it will be the largest of any land animal's—the size of a football and weighing from nine to twelve pounds. Numerous studies have compared the capacity of the elephant brain with that of human beings. For example, its temporal lobes, which store and retrieve sensory information, are large and complex relative to the size of the brain—denser and more convoluted than our own, in fact. The hippocampus, the primary seat of memory, is also large and well developed. Demonstrating an impressive memory, elephants forage over wide areas and varied terrain that can be altered by unpredictable droughts a decade apart, and they organize fairly complex groups with an array of vocalizations. Clearly they recall foraging routes; location of water sources and shelter; and, of course, extensive and ever shifting social networks. Tests demonstrate that they can process large numbers both of vocal tones and visual symbols. Perhaps it's true that elephants never forget.

Few muscles work harder than an elephant's heart.

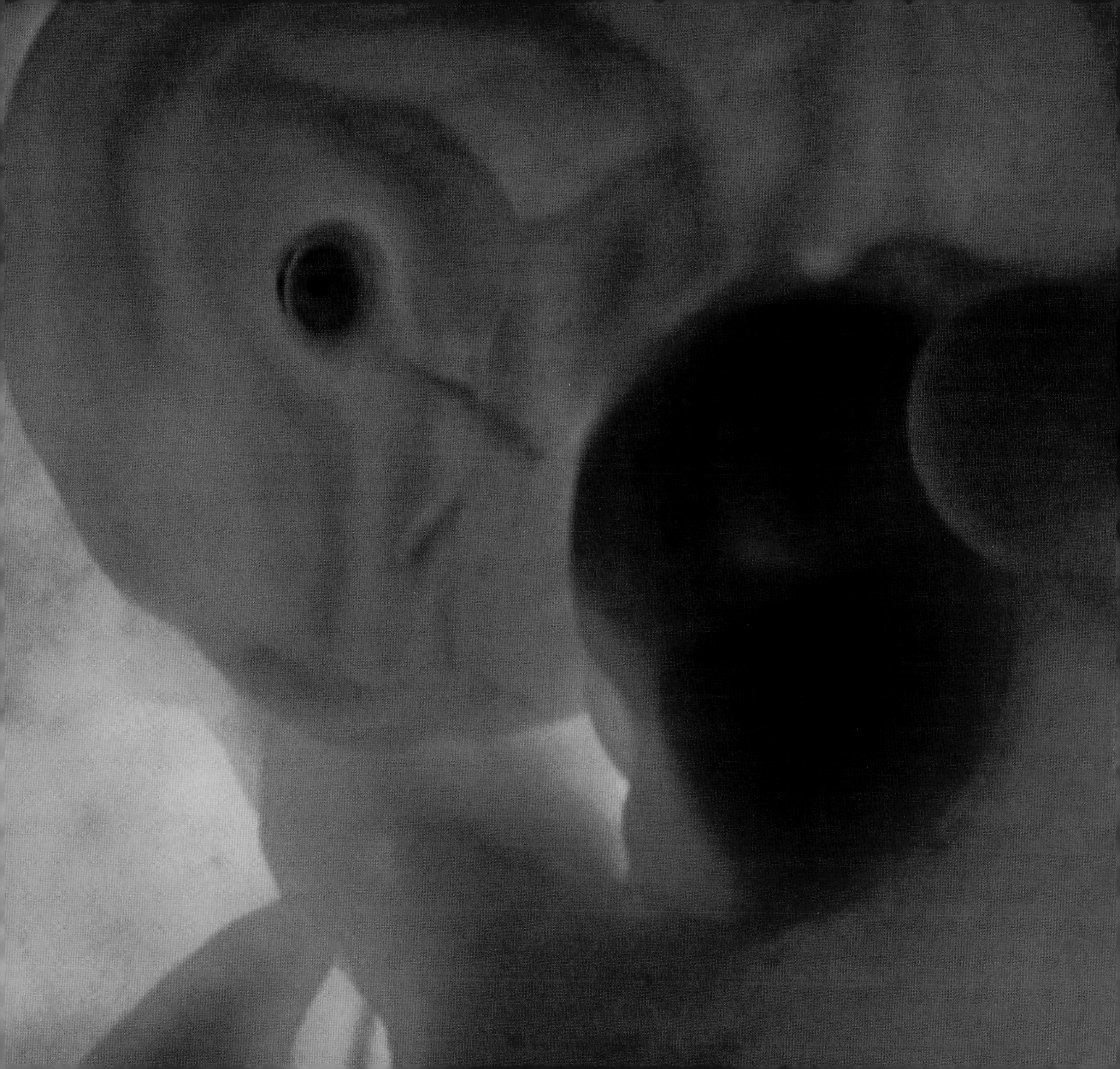

Night of the Living Dead

In our tour of extreme adaptations to the needs of reproduction, it would be difficult to find more outrageous adaptations than those demonstrated every day by two creatures whose gestation and birth we will peek into for comparison as we follow the elephant's progress—a lemon shark and a parasitic wasp. As everyone from biologists to comedians have pointed out, reproduction seems to be nature's main preoccupation. But to keep things lively, every species gets to make babies in its own crazy way.

The lemon shark, *Negaprion brevirostris,* is so named because of its brownish yellow skin color. It lives mostly in the tropics and subtropics, in the waters along both the Atlantic and Pacific coasts of North America, as well as down the Ivory Coast of Africa. It frequents sheltered shallow water around mangroves and coral reefs. Usually these heavy, blunt-nosed sharks reach a length of no more than eight or ten feet, although at least one 12-footer is on record, but all of them are large enough to never have to worry about predators themselves. The ecology of the lemon shark is better known than that of any of its kin, because it survives well in captivity. Although its iconic shape and ferocious maw strike terror in people who see it either live or on TV, there have been no fatalities among the handful of unprovoked attacks on record.

Not surprisingly, however, such an efficient predator leads a violent personal life. "The battle of the sexes" is very real for sharks. Soon after the female releases a chemical signal into the water, a male arrives in hot pursuit. Although probably they've never met before, they don't take time to get to know each other. His idea of foreplay is to immediately chase and bite the female, but fortunately she is literally thick skinned; sometimes a female's skin is twice as thick as a male's. At best, she will emerge from this bout with dramatic scars, and some females don't survive mating at all. Yet most not only live through it with one male but actually proceed to mate with others, bringing the largest possible genetic diversity to their investment in the future.

The male has no penis. He makes up for it with claspers, two symmetrical appendages formed by modifications of the ventral fin and permanently stiffened with cartilage. These appendages are visible on the underside of the body, making gender identification easier with sharks than with, say, dolphins. To mate, the male grasps the female, turns her sideways, and inserts one of his claspers. They may mate for as long as 20 minutes; unlike dolphins, they don't have to hurry while watching for trouble. Who would dare interrupt?

As you might imagine, mating without a penis requires some fancy engineering. Just as dolphins lock together in watertight intercourse, so have sharks evolved insurance against the risks of mating in salt water. While the male was swimming, he was holding the claspers in a fixed position to siphon seawater into a nearby sac; he can also flex the claspers to perform the same action. Dormant sperm enter the water-filled sac, protected by secretions that coagulate upon contact with seawater, creating sticky sperm bullets the size of shotgun pellets. As the male and female mate, the sac's compressor muscles contract and discharge the sperm packages into the female's vagina. Here the guardian

The parasitic wasp deposits eggs inside caterpillars' bodies.

pellet of protein dissolves. The dormant sperm sense female hormones in this environment and become active—competing with their fellows and with the sperm of other sharks that they encounter during their race to the egg.

Many people joke that shark movies keep them out of the ocean, but learning about parasitic wasps may scare you out of your own backyard. With over 60,000 species worldwide and more than 3,000 of those in the United States alone, one of their horrifying clan is probably near you at this very moment. Often they are recognizable by their elongated abdomen and, in many females, an ovipositor that is even longer than the stretched-looking body.

One particular species, *Cotesia glomerata,* long established in Europe and elsewhere, was introduced into the United States in 1883, in the hope that it would reduce the damage caused by another involuntary immigrant, the cabbage worm, the caterpillar of the cabbage white butterfly. You can see these smallish dusty-looking white butterflies sipping dandelion nectar or loitering suspiciously around your broccoli shoots. They devour cole crops, including cabbage, broccoli, and cauliflower. The butterfly, in turn, foolishly had been introduced into Canada in the 1850s and spread quickly across the continent. In Europe, northern Africa, southern Asia, and North America, *glomerata* is definitely the chief worry in the life of a cabbage worm.

Parasitic wasps blur definitions of predator and parasite. Usually we define a predator as a carnivore that pursues, kills, and eats another animal, and a parasite as a creature that lives off another and harms it in some way. Most of the time, a parasite can't afford to kill its host. But these wasps find a way around the rules. Where a penguin would invest vast resources in a single egg, or an elephant would carry her young for almost two years, *glomerata* turns to a more selfish method of providing for offspring.

First she locates her prey by following the subtle odor of plant damage, the silent alarm broadcast by many plants as they are eaten by herbivores or otherwise harmed. Here the wasp finds very young caterpillars—and attacks. Hardly defenseless, they fight back, biting and injecting the wasp with a paralyzing neurotoxin. It slows the wasp but doesn't stop her. As she grabs a caterpillar, she changes from predator to parasite, thrusting her long, sharp ovipositor through its skin and quickly pumping eggs into the relatively empty area between its gut and its skin. This cavity is filled with hemolymph, the arthropod version of blood that surrounds the vital organs. Rich in nutrients, it will serve the caterpillar eggs as a fertile environment for growth.

The caterpillar's immune system fights back—or tries to. But the parasitic wasp is the only known animal that has evolved a kind

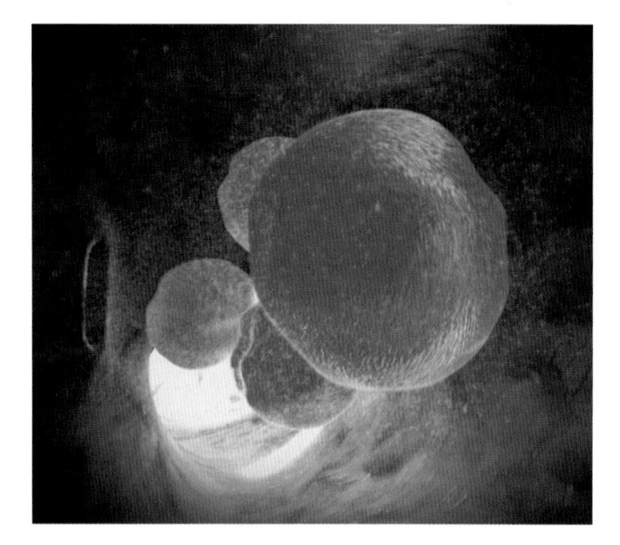

The wasp eggs invade the body of the caterpillar.

of symbiotic relationship with a virus that lurks in its own DNA. Inside the wasp's ovary, special cells coat the eggs with the virus, which disables the caterpillar's immune system, working much like the human HIV virus. The caterpillar loses the ability to recognize the wasp eggs as invading foreign bodies. It stops fighting them—which, for the caterpillar, is the beginning of the end. Next the virus hacks into the caterpillar's sex hormones and defuses them, chemically castrating it so that it will cease the maturation process.

The wasp moves relentlessly from one caterpillar to the next, injecting each with up to 30 eggs. The caterpillars are now merely zombies, surrogate wombs for the wasps that will grow inside them. The wasp eggs will hatch inside the bodies of the caterpillars. Instinctively resisting the temptation to eat their hosts' vital organs, the wasp larvae will gorge on caterpillar blood and grow ever larger over the next couple of weeks. Then, dripping an acid-like enzyme from their jaws, they will eat their way out of the caterpillars' paralyzed bodies.

This method of reproduction seems so horrific, so merciless, that it disturbed even Charles Darwin. "I cannot persuade myself," he sighed in 1860, "that a benevolent and omnipotent God would have designedly created the [parasitic wasp] with the express intention of their feeding within the living bodies of Caterpillars." More than a century later, this horrifying method of reproduction inspired nightmares in one of the scriptwriters of the 1979 horror film *Alien,* leading eventually to the now classic scene of the alien's offspring bursting out of an astronaut's chest.

A close-up of a shark's clasper

Snorkel Month 4

Now the embryo is beginning to look like an elephant. Admittedly, it's still a simplified cartoon elephant, but soon there will be no mistaking the appendage that prompted the name for this venerable order of animals. Elephants and their extinct cousins, such as mammoths and mastodons, are grouped together as members of the order Proboscidea, which simply means "possessing a proboscis," a nose. Elephants possess such an impressive nose that it is their identifying feature, like a zebra's designer stripes or the ridiculous neck of a giraffe. Finally, after about four months of gestation, this elephant embryo is growing a trunk.

It began much like any other mammalian proboscis, as a modest bulge. But now it is elongating like Pinocchio's nose. Although the embryo has 18 months left before birth, the trunk will need it; it takes a long time to build something so complex. In adulthood, the trunk will be strong enough to carry trees but a hundred times more sensitive than our fingertips. To accomplish the elephant's daily to-do list, the trunk requires 40 *thousand* individual muscles. The entire human body, by contrast, contains only 639.

This early appearance of the trunk is significant. Scientists argue that structures that differentiate and begin developing their specialties so early in the game date from the animal's distant past. Fossils, the structure of living animals, biochemical analyses—all confirm that elephants' nearest living relatives are not moose or any other large land mammals, but are actually manatees and dugongs.

At first glance, these aquatic mammals seem a long way across the family tree from *Elephas maximus.* Yet biologists think that elephants descended from aquatic ancestors. At this early stage, deep inside its developing kidneys, the elephant fetus is growing primitive funnel-shaped ducts called nephrostomes. Although these structures are well known in aquatic animals such as frogs and freshwater fish, and show up in primitive semi-aquatic animals such as the platypus, they have never been recorded in any other mammal that gives birth.

Not surprisingly, these once marine animals are Olympic-level swimmers. Elephants have been observed swimming 16 miles without stopping, to get from one Indian Ocean island to another. When an elephant swims, however, it must breathe through its trunk. Shortchanged in the neck department, it can't hold its mouth above water, so it keeps it closed and raises the trunk like a snorkel. It also uses the trunk in this way when wading in deep water. More than two millennia ago, Aristotle described how elephants sedately walk across river bottoms. If you live in the right part of the world, you can still witness this phenomenon, although your view from the shore may consist of only the tip of a trunk progressing across the surface like a sea monster's neck. The trunk may have evolved first to perform this job.

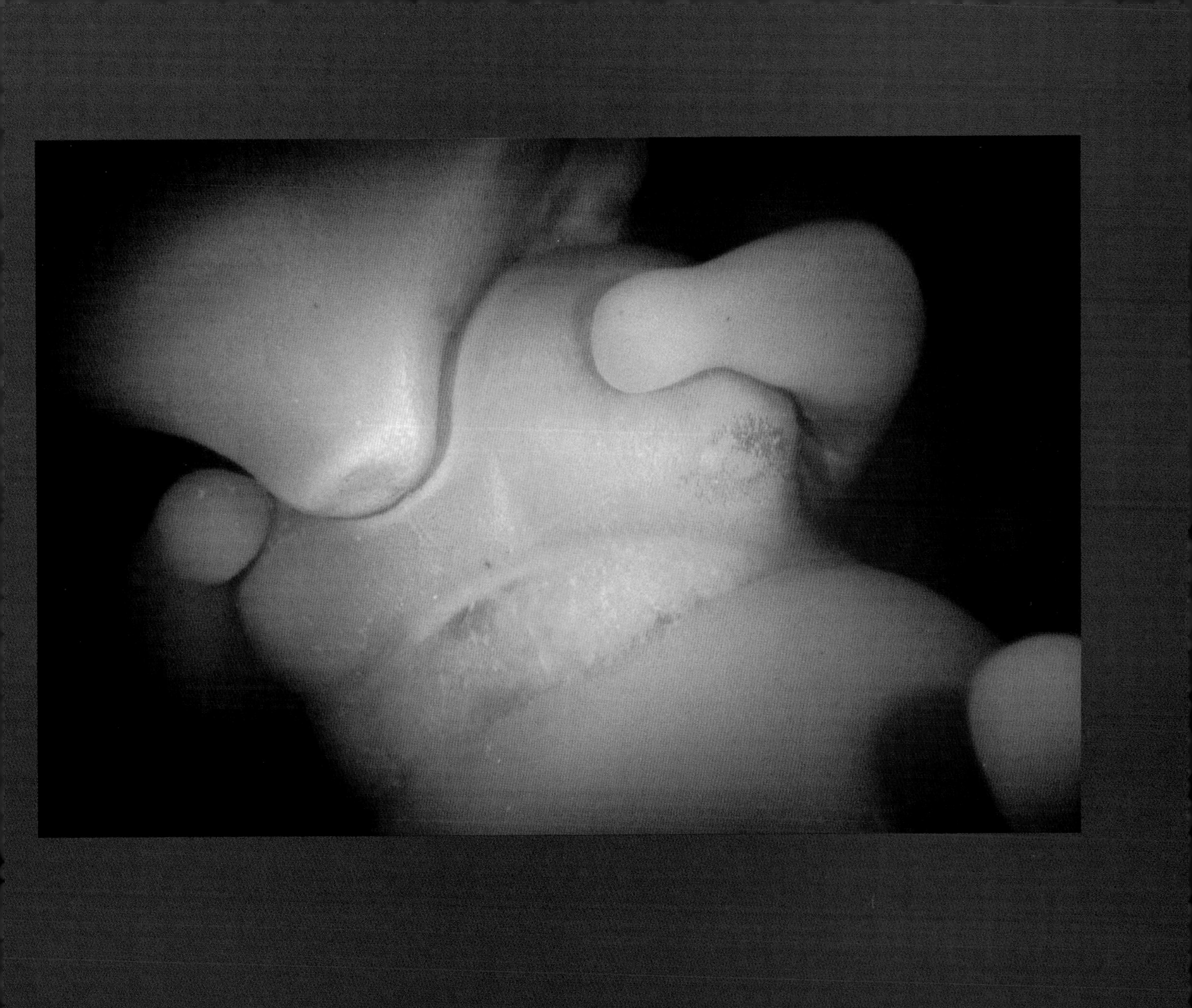

The elephant's trunk may have evolved in part to work as a snorkel.

A Walking Appetite Month 4

After four months in the womb, the elephant fetus is still only two inches long and weighs less than an ounce. Pink and covered in thin, unwrinkled skin, it still looks like a toy. The bright color comes from oversize blood vessels that are already developing. In adulthood, some of this elephant's veins and arteries will be more than 30 feet long, so their walls must be proportionately thicker to keep them from collapsing. Muscle cells and elastic fibers buttress the intersections between these far-flung circulatory highways.

The digestive system is also beginning to develop. Elephants evolved in a competitive world of plant-eaters. Although their ancestors had been evolving toward what we think of as an elephantine shape for millions of years, the Elephantidae family, which includes mammoths and mastodons, emerged as Africa's changing climate was spreading grasslands across the continent.

Proto-elephants had to compete with many herbivorous animals. Ruminants (cud-chewers) such as the antelopes were the dominant group in this environment, living on softer plant parts such as fruit and leaves, which are seasonal. While retaining a liking for fruit, early elephants evolved the ability to digest woody plant parts such as twigs and thorns, which are out of reach of many species and are also available throughout the year.

Thus elephants adapted to competition by growing ever larger. In general, larger animals have a slower metabolic rate, as in the relative heart rate of shrew and elephant (see page 126). Woody plant parts require a slow digestive system, because their nutrients need time to break down, with less than half of an elephant's diet fully digestible. So the evolutionary prompts to keep growing larger meant that they also had to spend more time foraging. Modern elephants devote almost every waking moment to eating, with an adult consuming up to 330 pounds of food daily (see page 116). By halfway through month four, the fetus is moving its legs in running motions. The head also swings slowly up and down, right and left. Such prenatal exercise helps the muscles develop properly and keeps the growing joints flexible. As elephants grew larger, they evolved thick, spongy pads on the soles of their feet, and these are now beginning to appear in the fetus. This fibrous tissue helps distribute weight and absorb shocks. Thanks to a convex bulge on the underside of the foot, the largest land animal can move in eerie silence. Even a twig underfoot doesn't snap audibly, because it is muffled under so much padding. Recent studies indicate that elephants also detect sound vibrations through their feet. Lower pitched vibrations—the calls of other elephants, thunderstorms, stampedes—travel much farther through ground than through air. Several distinctive adaptations, from specialized ear bones to nerves in the feet, indicate that elephants are picking up these vibrations.

Water's support permits whales to grow much larger than land animals. Every day a single blue whale consumes more than one million shrimplike shellfish called krill.

following pages: **An elephant fetus at 4 months after conception.**

The legs and digestive system are both adapted to serve a giant.

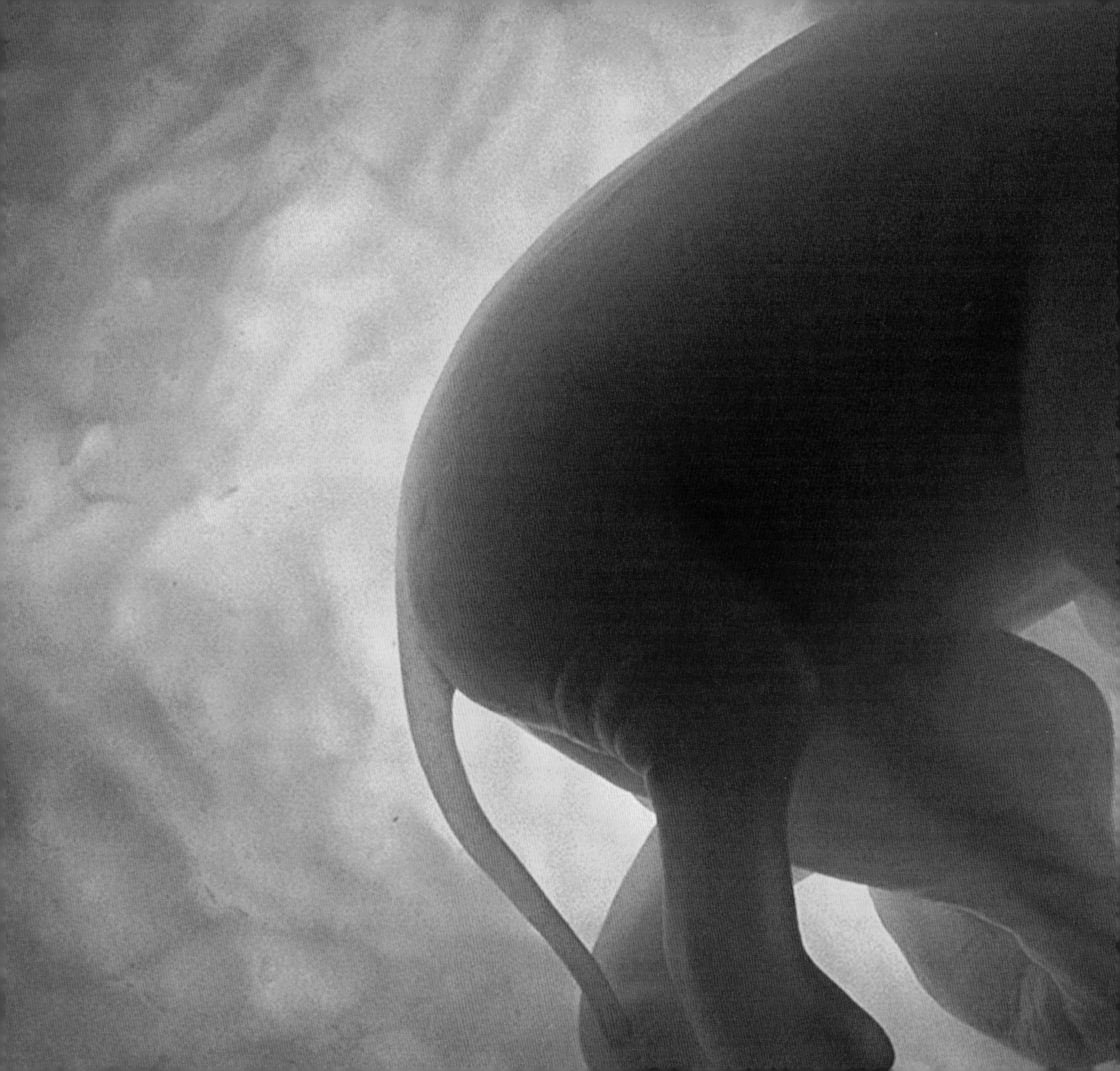

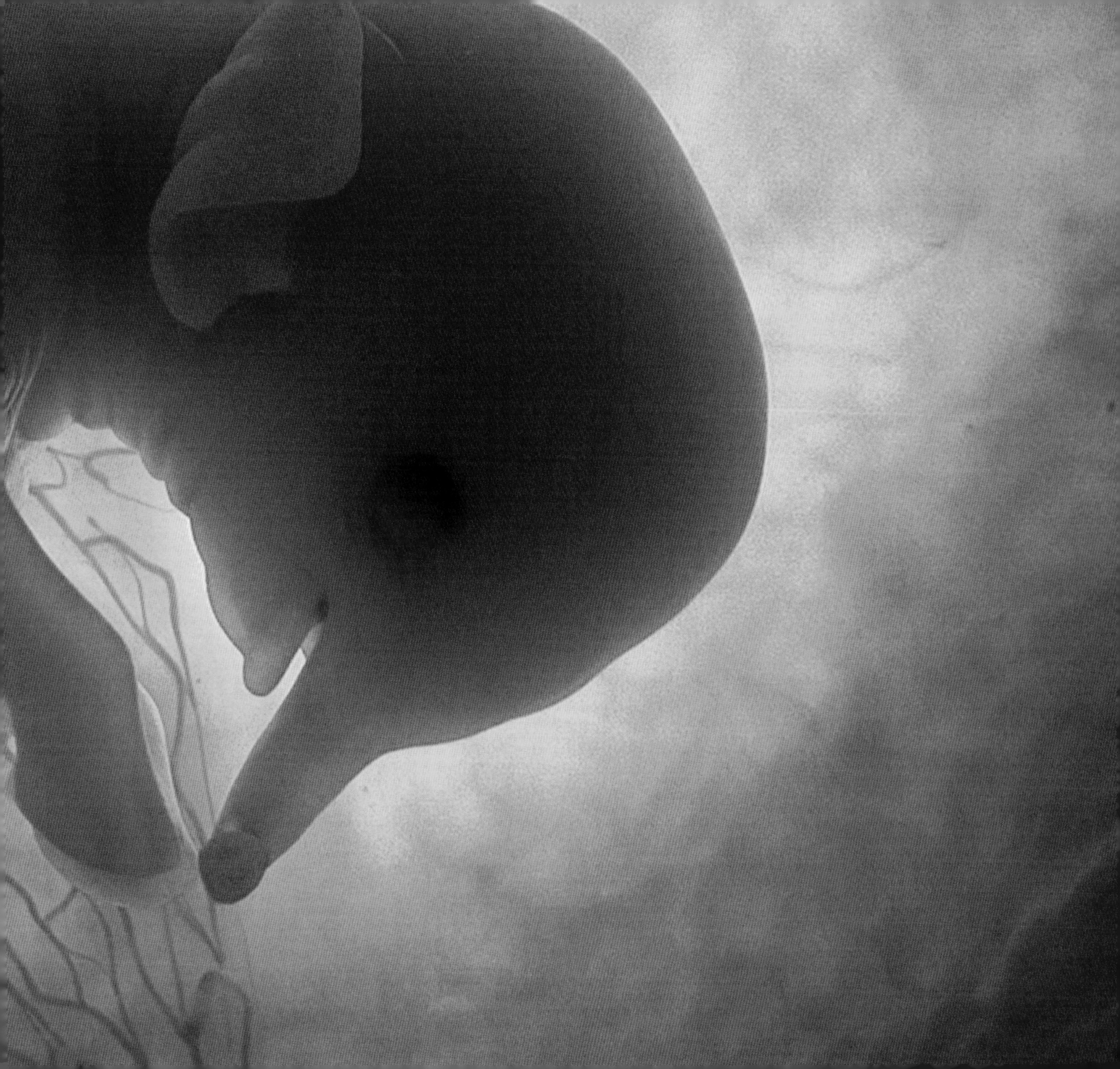

Suit of Armor

It has taken the more than one dozen shark pups almost a year to prepare for life outside the womb. A couple of weeks before birth, they are already about two feet long and have acquired their yellow-brown color. Only now does one of the lemon shark's lesser known adaptations begin to develop, during the fetus's last days in the womb. Despite their adult reputation as ferocious hunters, the young will be born relatively vulnerable to attack—from other predators, other sharks, and their own family. So in the past couple of weeks before birth, they grow a suit of armor.

Tiny razor-sharp scales, much smaller than the scales of fish, push through the surface of the skin. Biologists call them dermal denticles, which simply means "tiny skin teeth." The needlelike crowns of most denticles point toward the rear. If you rub a shark the wrong way, you will find that if you move your hand from tail toward head, the skin texture is sandpapery, while in the other direction it is relatively smooth. Different parts of the body reveal different lengths and angles of denticle—smoother and less ridged on areas that offer least resistance, such as the front edges of the fins and on the snout, and more sharply angled elsewhere. It's an impressive arrangement.

The aligned sharper ridges guide the flow of water around the shark's body, preventing the formation of eddying currents, thus reducing turbulence and drag—and allowing the shark to race through the water. Engineers and product designers often find their inspiration in the natural world, and swimsuit designers lately have begun offering suits designed on the sharkskin principle. They imitate the animal's hydrodynamic virtues by reducing drag at key points.

Sharks' denticles are indeed skin teeth, as you can see in the development of their heavily armed mouth. As terrifying to humans as the fangs of a snake or the claws of a tiger, sharks' teeth are enormously efficient but wear down quickly. Every week or so, a lemon shark will wear out a full set and replace them all. This is possible because sharks' teeth are actually scales, tougher and larger than skin scales, growing from and attached to the skin instead of anchored like ours in the jawbone. They replace their teeth as casually as we grow new hairs in place of those that fall out. As a consequence, their mouth always seems to have just been issued a new set of daggers. As many animals in the ocean have occasion to learn, sharks never run out of teeth and never go out to hunt unarmed.

Sharks' overlapping scales, like roof tiles or chain mail, grow at differing rates in different seasons, resulting in scale patterns like the rings that show age in a tree.

A shark's skin scales overlap like a coat of mail.

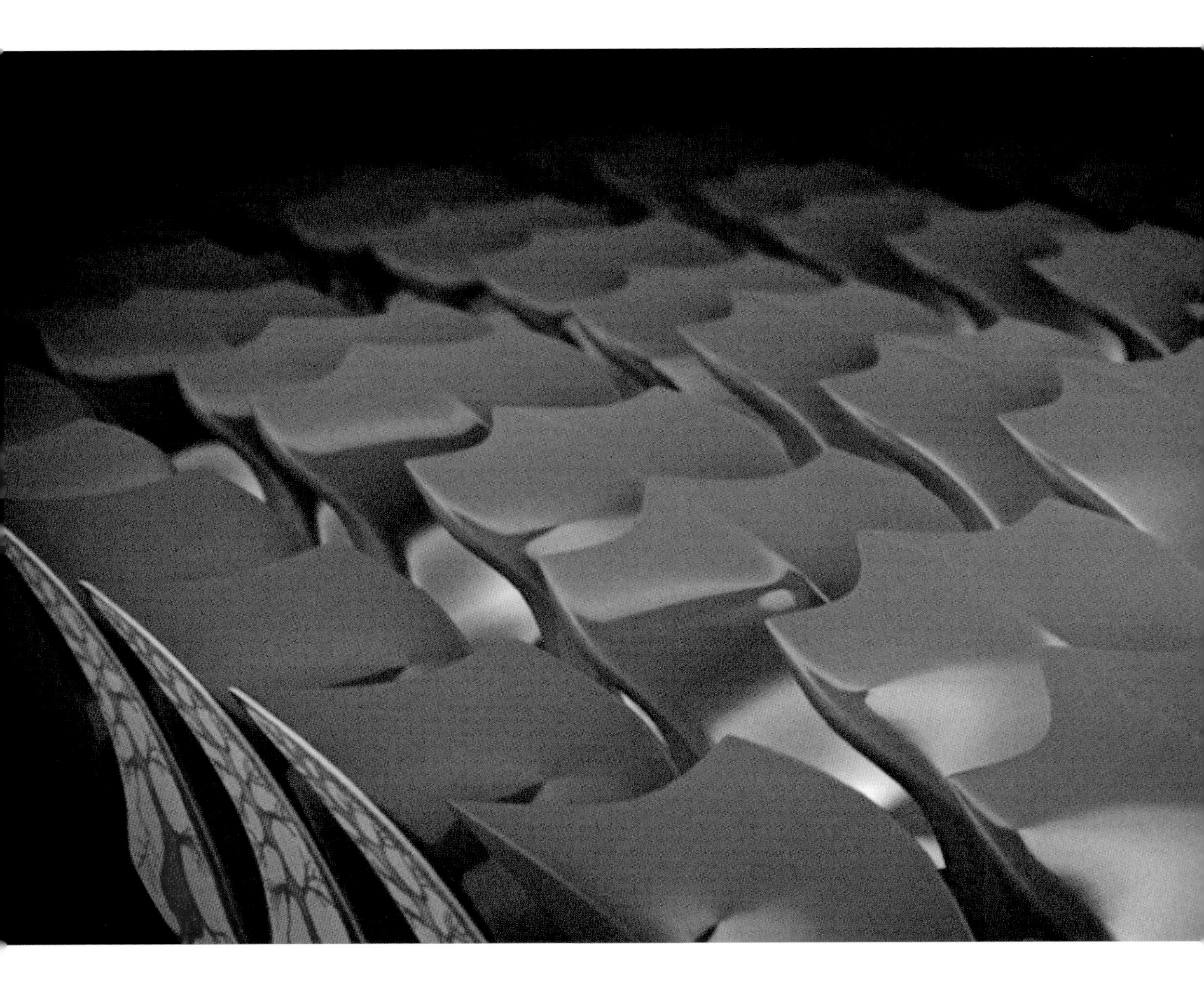

Domestic Violence

As if animals didn't have enough other problems to worry about from the moment of birth—from predation and food shortage to parasites and climate—many of them also must flee members of their own family. Nature is indeed "red in tooth and claw," as the poet Alfred Lord Tennyson described it, and often the violence begins at home. Numerous species, including human beings, practice infanticide—adult animals killing their own offspring or those of another member of the same species. There are various evolutionary rationales for this horrifying scenario. Consider, for example, the Indian monkey called the gray langur. Sometimes a rival male will overthrow a langur group's dominant male and then kill his mate's offspring so that she will reenter estrus and mate with the new boss. Endlessly creative nature added another subplot to this melodrama when female langurs evolved a phantom estrus that deceives the new male.

As strange and heartless as infanticide seems to us, it is matched by another widespread phenomenon—siblicide, in which a young animal kills one or more of its own brothers or sisters. Scientists recognize two kinds of siblicide. The rarer version is usually fatal to one or more participants. In the more common kind, the level of aggression within a litter or brood depends upon environmental factors such as scarcity of available milk or other resources. Siblicide has been documented in several species—arctic foxes, domesticated pigs, pronghorn, spotted hyenas, and, of course, human beings.

But a cousin of the lemon shark, the sand tiger shark, practices an even scarier version of siblicide. Sand tigers are those nightmarish sharks with rows of jagged teeth that project in various directions even when the mouth is closed. Like fish, sand tiger sharks have a rudimentary swim bladder, which provides them with greater buoyancy than other sharks, and they are also the only shark known to surface and gulp air, which they store in the stomach to further improve their ability to float motionless underwater. Surprisingly, they are relatively docile animals known to bother humans only when they have been bothered first.

Sand tiger females produce a new embryo every couple of weeks. This graduated age development going on in the same womb means that the eldest sibling grows tiny razor-sharp teeth well ahead of the next youngster in line. We're used to thinking of adult sharks as violent, but sand tigers are vicious even before birth. While still in the womb, pink and only four inches long, blind but already equipped with their clan's legendary scent-detecting abilities, the fetuses turn their alien-eyed, oversize head on each other and fight to the death—the death of all but one.

The fetal sand tiger shark will fight to the death in the womb.

Two Kinds of Nose Month 12

The next generation of elephant has spent a year in the womb and is just barely past halfway to birth. For ten more months, it will continue growing the specialized resources needed to operate the largest body of any land animal. Right now, however, it is only about 17 inches long and weighs only 26 pounds.

> "In the High and Far-Off Times the Elephant, O Best Beloved, had no trunk. He had only a blackish, bulgy nose, as big as a boot, that he could wriggle about from side to side; but he couldn't pick up things with it."
>
> **Rudyard Kipling, *Just-So Stories***

Its miniature legs are kicking against the wall of the uterus. Busy as they are, the legs seem undersize next to the more fully developed trunk. As we saw earlier, because of its complexity, the trunk takes a long time to develop, but already the fetus can move it in the womb, curling it over its head or up into its mouth. This is good practice. Elephants drink by siphoning up water and pouring it a trunkful at a time down the throat. Surprisingly, this awkward-sounding method can enable them to drink more than 50 gallons of water a day, and sometimes they have been observed doing so in less than five minutes.

In adulthood the trunk will, by itself, weigh more than two grown men. It will serve as drinking straw, trumpet, duster, tool, and weapon, but it will not be very impressive as a smelling organ. The elephant will be able to detect particularly strong odors with it, which is why elephants sniff people's shoes with the trunk. Subtle scents, however, would get lost in the long distance from trunk tip to brain cells. The nostrils lack olfactory cells in their epithelium, the wet lining of internal body cavities. The trunk is an extended nose, but it has lost much of its noselike function.

So the fetus is already growing a second kind of nose, a more sensitive organ for detecting smells. Like golden retrievers and all other mammals, including human beings, Asian elephants possess a vomeronasal organ, but theirs is the largest and most sensitive on Earth (see page 44). To analyze—or perhaps merely to enjoy—a more delicate aroma, it will use its trunk to gather a trace of the substance. Then it can insert the tip of the trunk in the roof of its mouth, where the vomeronasal organ is situated above a small opening in the palate, and let the trunk actually touch the duct that leads to it. This region is very close to the brain. Nerve endings in the vomeronasal organ translate the signal directly to the brain for quick and subtle analysis of the scent. The organ is so sensitive it can detect pheromones from more than a mile away.

The trunk is a nose, but a second smelling organ lurks above it.

Swim With the Sharks

A full year has passed since the female shark mated. Her more than a dozen eggs, each fertilized by a different father, have grown into embryos and fetuses and pups. Slowly, through a full circle of the seasons, they have journeyed toward birth, growing all the specialized adaptations they will need to become hydrodynamic, turbocharged fighting machines.

Extreme navigational and predatory senses are in place and ready to go. A shark's liver produces a lighter-than-water oil that keeps it buoyant. Its spine extends all the way down the tail to its tip, turning the whole body into a muscular propulsion system. Down sharks' sides run their famous lateral lines—narrow tubes equipped with neuromasts, minute hairs set in a gel that responds to even the tiniest vibration of movement or sound and conveys it to the hairs, which signal the nervous system. Pores on their snout, called the ampulae of Lorenzini, are ready to detect the almost immeasurably tiny electrical currents stirred by other animals' heart or brain or muscles.

The ancestors of sharks laid eggs shortly after conception, and small sharks still do today. But larger sharks have evolved proportionally larger pups that require longer development time in the womb. Yet the pups' yolk sac has not evolved to keep pace and provide all the necessary nutrients for a larger pup over a longer period of time. Instead, in an amazing transformation, the sac turns itself into a placenta. As it depletes its built-in resources, the sac attaches to the membranes on the wall of the uterus, sending out little filaments of blood vessels to connect with the mother's circulatory and digestive systems and feed off of her. Now, inside the womb, the pups thrash about, tearing the embryonic membrane sacs that still surround them. Some of this still nutritious material emerges from the mother's body ahead of her pups. Remoras—the sucker fish that hitch rides on sharks and turtles and even larger fish—rush to feast on it.

"There isn't any symbolism. The sea is the sea. The old man is an old man. . . . The shark are all sharks no better and no worse."

Ernest Hemingway, on *The Old Man and the Sea*

To give birth, the mother seeks a relatively secluded area away from predators that might eagerly dine on her young before they have even launched their life in the open sea. Not a sentimental mother, she might well eat them herself if not for the calm-inducing birth hormone that is now coursing through her system. Soon a tail emerges from her body. The pup is struggling to move as quickly as possible; he does the work while his mother slowly swims. He must pull hard to snap the umbilical cord and free himself. His siblings follow right behind him, 14 of them in under an hour—the next generation of what many people call the lord of the sea.

A newborn lemon shark, with umbilicus still attached

Hot and Cold Month 14

Many of the elephant fetus's stages of development concern an animal's ever present need to maintain body temperature. By month 13, for example, the elephant's penis is visible to ultrasound. No testicles show because they're inside the abdomen. This is an unusual arrangement but not unheard of; sloths and armadillos carry theirs within, as do anteaters. Humans and most other mammals have external testicles because sperm can't survive normal body temperatures, but the elephant's core body temperature is about half of one degree cooler than an average human's. Apparently this is enough of a difference to protect the sperm. Some scientists think that the internal testicles are further evidence of the elephant's aquatic past. As we saw with dolphins (see page 68), external testicles would interfere with a streamlined hydrodynamic shape. Whatever their ancestry, in adulthood his testicles will weigh around 13 pounds and produce five billion sperm per day.

Hairs are growing in a few key places by month 14. As they do on dogs, on elephants the important sensory hairs appear first. This fetus has eyelashes, a tail, and some hairs around its mouth. The rest of its body is completely bald but will soon grow the brownish red hair that contributes so much to the goofy charm of newborn elephants.

Again like dogs (see page 44), elephants have few sweat glands. Living in a tropical environment, however, they need ways to cool the oversize body. The large ears help. At the moment they are only four inches across, but they're destined for a two-foot span in adulthood. Their large size will help the elephant regulate its body temperature. Warm blood from throughout the body is carried to the ears, which the elephant fans back and forth to encourage cooling. It will even spray water on its ears to help them cool. Then the cooled blood returns to the body, lowering its temperature by as much as ten degrees Fahrenheit. The elephant's entire 100 gallons of blood can circulate through the ears' cooling system in about 20 minutes. The African elephant's ears are much larger even than these, because it lives in hotter and less shaded regions than the Asian does.

Living in the baking outback, kangaroos have evolved a cooling method similar to elephants'. Their forearms are covered with a dense network of blood vessels just beneath the surface. In hot weather, the kangaroo's blood pumps to the forearms, which the animal licks again and again to encourage heat evaporation similar to that achieved by panting. Surprisingly, Antarctic penguins are also likelier to overheat than chill down. Their interlocking layers of extra-thick feathers provide such great insulation that they can actually overheat when not in the water. They have to pump hot blood to their feet, where it can dissipate the warmth—and, when necessary, protect their egg or fledgling.

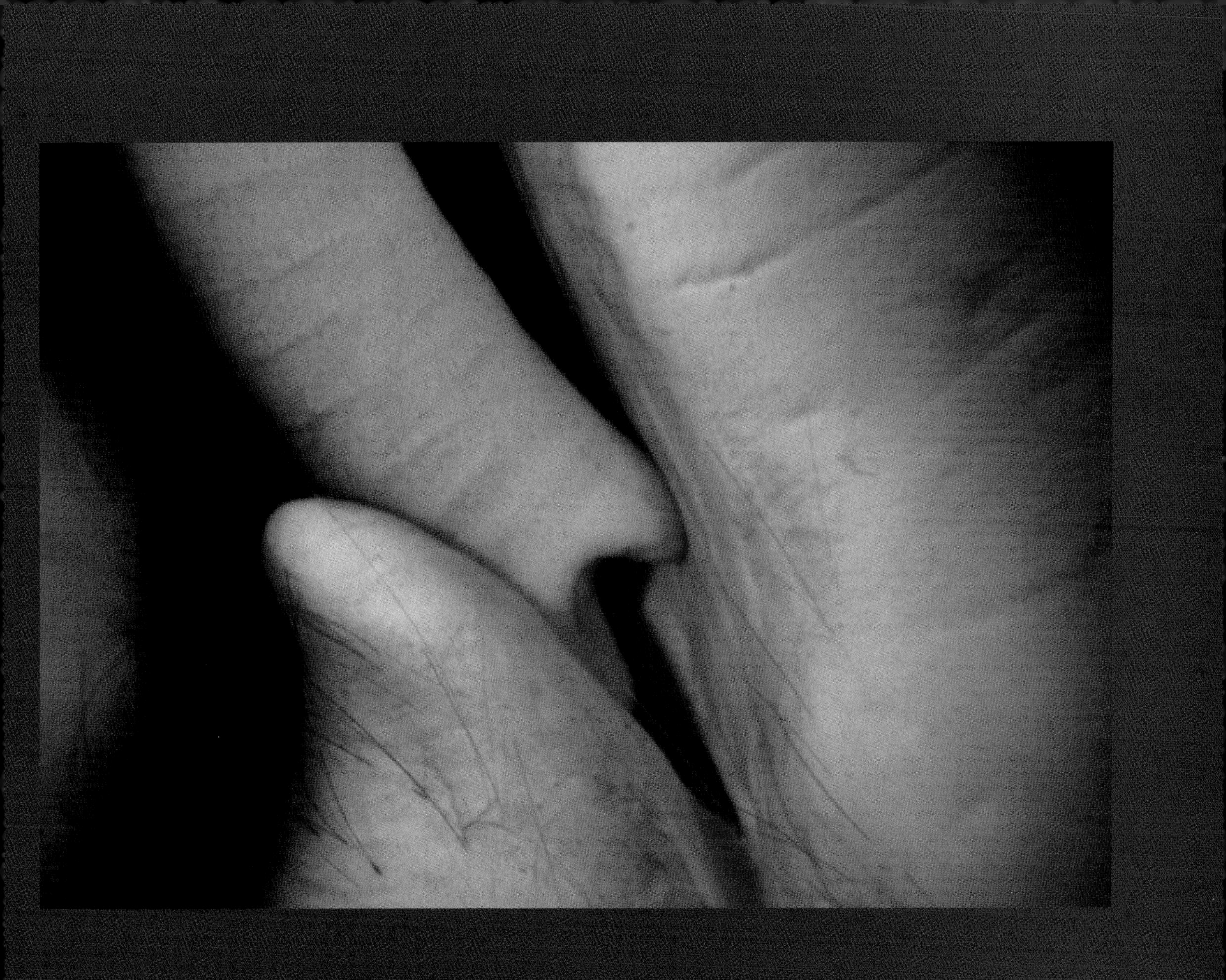

The fetus even practices drinking movements in the womb.

Coming Soon Month 19

There may be only one fetus in this elephant's womb—as opposed to the violent mob the shark was carrying—but the space is still getting crowded. With three months left before birth, both baby and mother are already uncomfortable. An elephant's giant stomach, which it requires for digesting bushels of coarse food, simply takes up too much space in the abdomen.

But still the fetus grows. Already at 140 pounds, it is adding a pound every day or so. It is still small enough that, if born prematurely, it would be unable to nurse because it wouldn't be tall enough to reach its mother's teats and would starve. So it now has one major job left—to keep growing. It will continue to do so throughout its life. Most human beings gradually shrink in height as they age after maturity, but elephants keep climbing. Their growth rate slows after sexual maturity, but it doesn't stop. The bigger the elephant, the older it is likely to be.

By this time, the fetus is covered in bristly reddish brown hairs that will fade away only after it reaches adulthood. Asian elephants have more hair than their African cousins, but it doesn't compare with that of most other terrestrial mammals. During its youth the hair will help protect its skin from mosquito bites and from the tropical sun, although Asian elephants in their natural environment have a more shaded habitat than do African.

Its gray skin is surprisingly delicate and thin and will take some time to thicken into the tough hide we associate with elephants. It will be so sensitive it will feel a mosquito land. The skin never becomes much thicker than a horse's, and never approaches the armored hide of a rhino. Shortly after birth, the calf will learn to cover himself with dust and mud as sunblock, and for some time he will be small enough to stand upright under his mother's belly, safe in the shade.

The skin is also acquiring its signature wrinkles, which is another useful adaptation. When he sprays water on himself, the countless nooks and crannies of his skin will trap it. Then, as the water evaporates, it will help lower the elephant's body temperature. Scientists estimate that this evaporation method meets about 75 percent of the elephant's heat loss requirements.

Of course, not all the fetus's changes are measured in increasing size. Its brain is still maturing as well, establishing the intricate networks of neural connections, the complex web of inherited species behavior and room for learning from experience that will enable it to take its place as a long-lived animal in a busy social group.

By 19 months, everything is growing toward the finish line.

A Breath of Air Birth

The longest journey in this book has reached its destination. The Asian elephant is ready to give birth. Over the course of almost two years, deep inside her body, one of the perennial wonders of life has been taking place—a single cell slowly turning itself into an individual animal. The calf will now leave this refuge and play his own role in the future of his species.

All sorts of preparations have been taking place inside the mother's body. Sometime during the past few days before birth, progesterone levels drop. Ever since conception, for 22 months now, levels have remained high. When dealing with a pregnant elephant, zoo officials monitor progesterone blood levels closely, so we know that a sudden drop of as much as 50 percent followed by a return to normal means that birth is imminent. This steroid hormone, which occurs naturally in all mammals but in wildly differing amounts, is crucial in building embryos, ordering the menstrual cycle, and maintaining the complex network of activities that pregnancy comprises.

Usually the umbilical cord is no more than three feet long, so unlike some animals this elephant can't strangle on his own lifeline. Because it is so short, however, it can't remain attached during the birth process. A human must sever its umbilical cord after birth, and a chimpanzee mother is sometimes seen carrying her baby around briefly with the cord still linking them. But as the elephant moves down the birth canal, his cord breaks off at his abdomen, leaving his navel as a commemorative scar like a badge of his membership in the ancient and populous family of mammals. Most of the cord withdraws back into the uterus.

Although the baby elephant isn't out in the air yet, this break with his mother's body leaves him without oxygen. Yet he doesn't try to breathe. Again behaving like an aquatic rather than a terrestrial mammal, he holds his breath as if diving. He can hold it for ten minutes, or sometimes even twice as long, surviving with neither air from outside nor blood-dissolved oxygen from his mother. This ability may be related to another curious fact: Elephants are born able to swim immediately.

The mother is uncomfortable. Like many other elephants about to give birth, she has lost her appetite—quite a loss for an animal of her size. She has begun to stretch and kick randomly. Even from outside, looking at her abdomen, we can see the baby moving. Throughout the pregnancy, the elephant's fetal position has kept him facing toward his mother's tail. Now the womb's administrative chemistry issues a silent command to turn him around and aim his back feet toward the birth canal. He is upside down now, on his back, with his legs held upward, as he starts to pass through the cervix.

Labor takes about an hour. The first glimpse from outside is a bulge under the tail—the amniotic sac. Most elephants are born hind feet first, with the head visible only late in the process. Because many terrestrial mammals enter the world the other way around, headfirst, biologists speculate that elephants' method may be yet another clue to their aquatic ancestry. As we have seen, dolphins and sharks give birth as elephants do—rear end first, the

The first sign of birth is a glimpse of the amniotic sac.

tail emerging well before the wriggling form finally pulls its head out into the water (see pages 104 and 144). Dolphins, which must breathe air from the surface, can't afford to be born headfirst. If a birthing problem leaves the newborn stuck with his head in the water, he will drown. Porpoises, whales, manatees, and dugongs all follow the same rear-first procedure that this elephant employs.

Both the elephant calf's body and his brain will have a lot of growing left to do. At birth, his brain weighs as much as an adult's, but it has only begun its long process of growth and development. Like human beings, elephants arrive in the world with the brain structure established but many of its networks and processes still to be worked out over time. But elephants don't take quite as long for the brain to fully mature as we do; they manage it in 12 years or so.

Now the calf emerges into the world and takes his first breath. After only half an hour, he is rising on wobbly legs and looking around. He has trouble controlling his limbs and stumbles often. No matter how awkward he is at first, within the first couple of hours of his life he will begin a lifetime of walking, exploring his world. At "only" 270 pounds, he has the cartoonish disproportion that triggers our instantaneous fondness for baby mammals, including our own—oversize eyes, rounded body, short legs (see page 32). His vulnerable, innocent look helps explain why baby elephants are always a popular attraction at zoos.

This unfinished air is not an illusion. Apparently elephants are born, for example, with no idea how to use their dangling, oversize trunk. Many calves curl the tip of the trunk into their mouth and suck it like a baby sucking its thumb. Within a week, he will be using his trunk to pick up objects, but it will take him much longer to learn how to drink with it, and it will be years before he masters snorkeling.

At birth this baby can't see well. Most wild elephants are born during the night, or actually during the dark hours of early morning. Unlike kangaroo joeys, he is born without instinctive guidance to his mother's milk, and searches randomly around her abdomen, sometimes at the hind legs and sometimes at the front. His mother helps guide her near-blind calf to her teats so that he may suckle immediately. His eyes adjust slowly over the next couple of hours, as he peers around, blinking. They will be working properly before sunrise shows him his new world.

The newborn charms both mother elephants and zoogoers.

And Multiply Afterword

Now that the play is over, let's bring our characters back for a curtain call. We need to take a moment to look back at the several animals we have followed during the course of this book.

Out in the real world, each species mated during a particular time of year and progressed through its reproductive cycle as the seasons unfolded around it. For our purposes here, however, let's imagine that all three of our main characters—dog, dolphin, and elephant—mated on the same day. So did the penguin and the kangaroo and the shark. Meanwhile the parasitic wasp was locating caterpillars and depositing her eggs inside them. All of these events took place 22 months ago. Where are these other animals now, at the end of our journeys with them, the slowest-moving of them all, the Asian elephant, having finally given birth? (To see this progression all at once, turn to pages 16-17.)

The wasps were the first to reach maturity, racing off to inspire more horror movies. Only four days after their mother injected eggs into each caterpillar's body, they hatched into larvae. After less than a month of zombie buffet, the larvae ate their way out of each caterpillar's body. Many were snapped up by predators, but hundreds survived to pupate in cocoons. Ten months ago, the next generation emerged from their cocoons and went in search of caterpillars on which they could practice their peculiar version of maternal commitment. This generation will soon be ready to reproduce.

Twenty-one months ago, after only a month in the womb, the tiny kangaroo joey was born and crawled up to his mother's pouch to lock onto a teat. At about the age of six months, he began to leave the pouch and explore. A couple of months later, he could take care of himself. Females reach maturity a bit earlier, but our male is only just now looking for a mate.

The golden retrievers were the next to emerge from the womb, 20 months ago. A year is long enough for them to reach sexual maturity and conceive litters of their own, and many mature sooner. Nine pups were a high number for this breed, however, so let's say that two months after they mated each young dog produced seven pups. We would now have 63 pups that are already half a year old—almost ready to join in the mating game themselves. Some breeds would have grandchildren by now.

Just as the dogs were born, the penguin hatched—a breakthrough that required three exhausting days of chipping away at the thick shell. For the next five months, the parents took turns feeding the fuzzy chick and warming it on their feet. Then it went off for its first dive and began its independent life. The next spring, a year after its own conception, it was back and mating and starting the cycle all over again. Seven months ago, its own egg hatched, and the new chick has already grown into its adult tuxedo.

Ten months ago, after a year in the womb, the shark gave birth to 20 pups. Fortunately a birthing hormone anesthetized her usually ravenous appetite, or she would have gobbled up her own young. They won't reach sexual maturity for another 11 years or more. At the same time as the shark, the dolphin was born. A female might live half a century, and he has at least three decades ahead of him. They mature at different rates, but it will be a few years before this dolphin makes his own contribution to his species' future.

> "The very act of understanding is a celebration of joining, merging, even if on a very modest scale, with the magnificence of the Cosmos."
>
> **Carl Sagan**

And now we have the youngest of our youngsters, the elephant calf that was born today. Some elephants bear twins, but this one was born alone. He may suckle from two to three years, and he won't reach sexual maturity until he's 15. By the time this slow-motion elephant grows into a giant and fathers his first offspring, the other animals from this book will have scattered across their habitats. Many will have fallen to predators and disease, but many others will have survived and reproduced. The dog's nine pups will have died of old age, having produced hundreds more pups that soon began reproducing on their own.

I think my favorite aspect of these varied journeys is the reminder that nature often seems to think outside its own box. Consider some of the examples we've seen of evolution's unpredictability. When a shark's yolk sac finally depletes, it turns into a placenta. The kangaroo's milk keeps changing its nutrients to suit each stage of the joey's growth in the pouch. Elephants hear through the ground, and sharks detect through water the electrical impulses in the body of their prey. Dolphins descended from land animals, and elephants from marine animals.

All over the planet, human beings are amassing such facts and doing the best they can to interpret them. When nature reminds us that our explanations are approximate, we tweak them again. Taxonomists, for example, must periodically revise their descriptions of animal groups, lumping what were formerly considered separate species into one species or splitting one into more, to better accommodate either the growing amount of information or the ongoing evolution of species in the wild.

The Italian poet Eugenio Montale called one of his collections *Provisional Conclusions,* a title that would fit well on any scientific study. Even after thousands of years of scientific exploration, we are still revising our view of the world as we cross new frontiers. How exciting to decipher old mysteries and discover new ones in previously uncharted territory—inside molecules, at the bottom of the sea, beyond our galaxy, and in the womb.

Further Reading

This list includes books and websites that were useful in researching this volume, as well as other sources that will guide your curiosity about the topics. It excludes technical papers in scientific journals, which are essential in researching details but prove of little interest to the general reader.

On Dogs

Budiansky, Stephen. *The Truth about Dogs.* Viking, 2000.

Coppinger, Raymond and Laura. *Dogs: A Startling New Understanding of Canine Origin, Behavior, and Evolution.* University of Chicago Press, 2002. A fascinating look at dogs both before and after they partnered with humans.

Fogle, Bruce. *Dog Owner's Manual.* Dorling-Kindersley, 2003. A handy and informative field guide to choosing and caring for a dog.

——. *The New Encyclopedia of the Dog.* Dorling-Kindersley, 2000. An excellent one-volume survey of all things canine.

Morris, Desmond. *Dogwatching.* Crown, 1987.

Ruvinksy, Anatoly, et. al. *The Genetics of the Dog.* Cabi, 2002. Fascinating detailed information, but too technical for the casual reader.

On Dolphins

Reynolds, John Elliott. *The Bottlenose Dolphin: Biology and Conservation.* University Press of Florida, 2000.

http://www.dolphins.org/. The website of the Dolphin Research Center, with many useful links.

http://www.sampla.org/. The website of the South African Marine Predator Lab, with great details about research projects.

On Penguins

Davis, Lloyd S., and M. Renner. *Penguins.* Poyser, 1995. A scientific survey of penguin species and natural history.

Williams, Tony D. *The Penguins.* Oxford University Press, 1995. An excellent detailed overview of penguins around the world.

http://www.penguinworld.com/. A friendly general introduction to the variety of penguins.

http://www.seaworld.org/infobooks/Penguins/home.html. General information about penguins.

On Kangaroos

Dawson, Terence J. *Kangaroos: Biology of the Largest Marsupials.* Cornell University Press, second edition, 1998. Detailed scientific background about the lives of all species of kangaroo.

http://www.iberianature.com/trivia/etymology_mammals.htm. This site tracks the etymology of names of mammals, leading into surprising connections and histories. Start with *kangaroo.*

http://kango.anu.edu.au/. The site of the Kangaroo Genome Project, at Australian National University.

On Sharks

Stafford-Deitsch, Jeremy. *Shark: A Photographer's Story.* Bramley Books, 1987. A lively first-person account of shark research and photography all over the world.

http://www.sharktrust.org/. The website of an international research and conservation organization, the Shark Trust.

On Elephants

Masson, Jeffrey Moussaieff, and Susan McCarthy. *When Elephants Weep: The Emotional Lives of Animals.* Delacorte, 1995.

Sukumar, R. *The Asian Elephant: Ecology and Management.* Cambridge University Press, 1993. Detailed and at times technical, but an encyclopedic source on these topics.

http://animals.nationalgeographic.com/animals/mammals/asian-elephant.html. General information with links to photos and other articles.

http://nationalzoo.si.edu/Animals/AsianElephants/. A site for the Smithsonian National Zoological Park, with general information and links, including Web cams.

http://www.sanparks.org/parks/kruger/elephants/about/evolution.php. A South African National Parks website.

On Reproduction in General

Bainbridge, David. *Making Babies: The Science of Pregnancy.* Harvard University Press, 2001.

Clutton-Brock, Juliet. *A Natural History of Domesticated Mammals.* British Museum, 1987. This excellent history includes not only dogs and cows but even elephants.

Forsyth, Adrian. A *Natural History of Sex: The Ecology and Evolution of Mating Behavior.* Scribner's, 1986. A fine introduction to this all-encompassing topic.

Hrdy, Sarah Blaffer. *Mother Nature: Maternal Instincts and How They Shape the Human Species.* Pantheon, 1999. Although she focuses on human beings, Hrdy naturally places us in a larger context in this fascinating volume.

Ings, Simon. *A Natural History of Seeing.* W. W. Norton, 2008. A splendid tour of the evolution and function of sight in humans and other animals.

Lewin, Roger. *The Smithsonian Looks at Evolution.* Smithsonian Museum, 1982. A good general overview of the topic.

Lewis, C. Day. "Florence: Works of Art," part 5 of *An Italian Visit*, in *The Complete Poems of C. Day Lewis.* Stanford University Press, 1992.

Milner, Richard. *Darwin's Universe: Evolution from A to Z.* University of California Press, 2009. The most readable and entertaining volume ever written on this topic.

Parker, Steve. *Animal Babies: A Habitat-by-Habitat Guide to How Wild Animals Grow.* Rodale, 1994. Exactly what it says in the subtitle; a good introduction and a way to help children explore the topic.

Pliny the Elder. *Natural History*, translated by John Healy. Penguin, 1991. The Roman encyclopedia by one of the most insatiably curious people in history.

Rodgers, Joann Ellison. *Sex: A Natural History.* Times/Henry Holt, 2001. A huge encyclopedic tour of the topic.

Rose, Kenneth David. *The Beginning of the Age of Mammals.* Johns Hopkins University, 2006. A scholarly volume but well written and informative.

Shepard, Paul. *The Others: How Animals Made Us Human.* Island Press, 1996. An elegant survey of our slowly evolving attitudes toward our fellow creatures.

Tudge, Colin. *The Variety of Life: A Survey and a Celebration of All the Creatures That Have Ever Lived.* Oxford University Press, 2000. A book as grand as its title, yet always lucid and enlightening.

Wolpert, Lewis. *The Triumph of the Embryo.* Oxford University Press, 1991. For the serious reader, but lucid and accessible.

Sagan, Carl. *Cosmos.* Random House, 1980. The companion to the momentous public television series, and a great introduction to thinking about the world and its creatures.

http://amnh.org/. The American Museum of Natural History site, with many useful links.

http://www.exploratorium.edu/. The Exploratorium, an interactive science museum in San Francisco, California.

http://www.flmnh.ufl.edu. The Florida Museum of Natural History, with many links and special interest in dolphins.

About the Author

Michael Sims is the author of several books about nature and science, including *Adam's Navel: A Natural and Cultural History of the Human Form*, which was a *New York Times* Notable Book and a *Library Journal* Best Science Book, and *Apollo's Fire: A Day on Earth in Nature and Imagination*, which National Public Radio chose as one of the best science books of 2007. He has also edited several literature collections, including most recently *The Penguin Book of Gaslight Crime*. His writing has appeared in *Orion*, *New Statesman*, *Gourmet*, the *Washington Post*, the *Chronicle of Higher Education*, *American Archaeology*, and many other periodicals in the United States and abroad. He lives in Greensburg, Pennsylvania.

Learn more at *www.michaelsimsbooks.com*.

Acknowledgments

My first debt is to the scientists whose hard work discovered these wonderful details about the world, as well as to the researchers, writers, special-effects wizards, and producers of the documentaries that inspired this series of books. Part of my pleasure in this project emerged from the opportunity to work with such a great team at National Geographic Books in Washington, D.C.: editor Garrett Brown, designer Cinda Rose, design assistant Al Morrow, and copy editor Judith Klein; as well as picture editor Julia Behar at RocketRights in London. Garrett has been invariably helpful as well as patient, and I also want to thank executive editor Barbara Brownell Grogan.

Thanks to my agent, Heide Lange, and her assistants Alex Cannon and Jennifer Linnan, at Sanford J. Greenburger Associates, Inc., in New York; to Bill and Rhonda Patterson, at whose North Carolina beach house I began work on this book; to Sarah Patterson, who was endlessly encouraging, as always; to the wonderful staff of the Greensburg Hempfield Area Library, especially Cesare Muccari, Diane Ciabattoni, Cindy Dull, and Linda Matey; and to the invaluable and computer-savvy Karissa Kilgore. Love and thanks to my wife, Laura Sloan Patterson, who encouraged and advised even while completing a book of her own.

I dedicate this book to my mother, Ruby Norris Sims, whose tolerance for countless pets helped inspire a lifelong fondness for animals.

In the Womb on the National Geographic Channel

This stunning series from Pioneer Productions opens a window into the hidden world of the fetus, and explores it in amazing new detail. Revolutionary 3-D and 4-D ultrasound imagery sheds light on the delicate, dark world as never before, both in animals and humans. Every journey with *In the Womb* is unique, and you'll find them all on National Geographic Channel.

Also available on DVD at shopNGvideos.com or 1-800-627-5162.

In the Womb television credits

In the Womb: Animals
© 2006 Pioneer Productions & Fox TV Studios
Written, Produced, and Directed by Yavar Abbas
Assistant Producer, Lizzie Abbott
Executive Producer, Simon Andreae

In the Womb: Extreme Animals
© 2008 Pioneer Productions and Fox Television Inc.
Written, Produced, and Directed by Peter Chinn
Assistant Producer, Hannah James
Executive Producer, Andrea Florence and Simon Andreae

In the Womb: Animals series
Special Effects, Artem and Bandito
Special Effects Director, David Barlow
Head of Development, Jeremy Dear
Head of Production, Kirstie McLure
Executive Producer for the National Geographic Channel, Jenny Apostol
Research Supervisor for National Geographic Channel, Genevieve Sexton

With thanks to Stuart Carter at Pioneer Productions for his dedication and commitment to the art of communicating science; to the production teams whose vision and expertise have brought these extraordinary stories to life; and to Dr. Thomas B. Hildebrandt, Dr. Jose I. Castro, Professor John Rodger, Dr. Hans Smid, and Professor Nick Sparks for their guidance and help in facilitating these films.

Picture Credits

2-3 (all): Pioneer Productions; 4-5: Pioneer Productions/David Barlow; 6: istock.com/Florea Ionut; 7: istock.com/Nicolas Delafraye; 8 (both): Pioneer Productions/David Barlow; 9: istock.com/Pauline S. Mills; 10: istock.com/Matt Jeacock; 11: Tony Polack; 12: AP Photo/Taiji Whale Museum via Kyodo News; 13: Michael Sims; 14: David Barlow Photography; 15: istock.com/Nikolay Suslov; 18: istock.com/Jeff Dalton; 18-19: istock.comMike Dabell; 19: istock.comMartesia Bezuidenhout; 20-21: istock.com/Andrew Howe; 23: istock.com; 25: Pioneer Productions; 27: Pioneer Productions; 31: Pioneer Productions; 33: Pioneer Productions; 35: Pioneer Productions; 38-39: Pioneer Productions/David Barlow; 41: Pioneer Productions/David Barlow; 43: istock.com/John Pitcher; 45: Pioneer Productions/David Barlow; 47 (all): Pioneer Productions; 49: Pioneer Productions/David Barlow; 51: Pioneer Productions/David Barlow; 53: Jason Stone/Woodfall Wild Images/Photoshot; 55: Pioneer Productions/David Barlow; 56-57: ardea.com/John Daniels; 59: istock.com/Virginia Hamrick; 60: istock.com/Jan Will; 60-61: istock.com; 61: istock/Anthony Brown; 62-63: ardea.com/Augusto Stanzani; 65: ardea.com/M. Watson; 67: istock/Lars Christensen; 69: ardea.com/Augusto Stanzani; 71: Pioneer Productions; 73: Pioneer Productions; 75: ardea.com/M Watson; 76: Pioneer Productions; 76-77: Pioneer Productions/David Barlow; 77: Pioneer Productions/David Barlow; 79: Oxford Scientific/Alan Root; 80-81 (all): Pioneer Productions; 83: Pioneer Productions/David Barlow; 84-85: Pioneer Productions/David Barlow; 87: istock/Centrill Media Corporation; 89: Pioneer Productions/David Barlow; 91: Pioneer Productions/David Barlow; 93: Science Photo Library/Thierry Berrod, Mona Lisa Production; 95: Science Photo Library/US Navy; 99: istock.com/Michele Lesser; 101: Pioneer Productions/David Barlow; 102: istock.com; 103: ardea.com/M. Watson; 105: ardea.com/Augusto Stanzani; 106-107: ardea.com/Augusto Stanzani; 109: ardea.com/Augusto Stanzani; 110: istock.com/Chris Dascher, 110-111: Pioneer Productions; 111: istock.com/Simone van den Berg; 112-113: ardea.com/Jagdeep Rajput; 115: ardea.com/Joanna Van Gruisen; 117: ardea.com/Jagdeep Rajput; 118-119 (both): Pioneer Productions; 121 top (both): Pioneer Productions; 121 left (both): Pioneer Productions/David Barlow; 121 right (both): Pioneer Productions; 123: Pioneer Productions/David Barlow; 125: Pioneer Productions/David Barlow; 127: Pioneer Productions/David Barlow; 129: Pioneer Productions; 130: Pioneer Productions; 131: Jeffrey C. Carrier / SeaPics.com; 133: Pioneer Productions/David Barlow; 135: Pioneer Productions/David Barlow; 136-137: Pioneer Productions/David Barlow; 139: Pioneer Productions; 141: Pioneer Productions/David Barlow; 143: Pioneer Productions/David Barlow; 145: Doug Perrine/SeaPics.com; 147: Pioneer Productions/David Barlow; 149: Pioneer Productions/David Barlow; 151: Pioneer Productions; 153: Pioneer Productions.

IN THE WOMB Animals

Michael Sims

Published by the National Geographic Society

John M. Fahey, Jr., President and Chief Executive Officer
Gilbert M. Grosvenor, Chairman of the Board
Tim T. Kelly, President, Global Media Group
John Q. Griffin, President, Publishing
Nina D. Hoffman, Executive Vice President; President, Book Publishing Group

Prepared by the Book Division

Kevin Mulroy, Senior Vice President and Publisher
Leah Bendavid-Val, Director of Photography Publishing and Illustrations
Marianne R. Koszorus, Director of Design
Barbara Brownell Grogan, Executive Editor
Elizabeth Newhouse, Director of Travel Publishing
Carl Mehler, Director of Maps

Staff for This Book

Garrett Brown, Editor
Judith Klein, Copy Editor
Julia Behar, Picture Editor
Julia Madonna, Picture Researcher
Cinda Rose, Art Director
Al Morrow, Design Assistant

Jennifer A. Thornton, Managing Editor
R. Gary Colbert, Production Director

Manufacturing and Quality Management

Christopher A. Liedel, Chief Financial Officer
Phillip L. Schlosser, Vice President
Chris Brown, Technical Director
Nicole Elliott, Manager
Monika D. Lynde, Manager
Rachel Faulise, Manager

Founded in 1888, the National Geographic Society is one of the largest nonprofit scientific and educational organizations in the world. It reaches more than 285 million people worldwide each month through its official journal, *National Geographic,* and its four other magazines; the National Geographic Channel; television documentaries; radio programs; films; books; videos and DVDs; maps; and interactive media. National Geographic has funded more than 8,000 scientific research projects and supports an education program combating geographic illiteracy.

For more information, please call 1-800-NGS LINE (647-5463) or write to the following address:

National Geographic Society
1145 17th Street N.W.
Washington, D.C. 20036-4688 U.S.A.

Visit us online at www.nationalgeographic.com

For information about special discounts for bulk purchases, please contact National Geographic Books Special Sales: ngspecsales@ngs.org

For rights or permissions inquiries, please contact National Geographic Books Subsidiary Rights: ngbookrights@ngs.org

ISBN: 978-1-4262-0175-2

Library of Congress Catalog-in-Publication Data
Sims, Michael, 1958-
In the womb : animals / by Michael Sims.
p. cm.
Based on two television documentaries about animals in the womb.
Includes bibliographical references and index.
ISBN 978-1-4262-0175-2 (hardcover : alk. paper)
1. Embryology--Pictorial works. 2. Fetus--Development--Pictorial works. I. Title.
QL956.S56 2009
571.8'61--dc22

2008049996

Printed in China